FOOD SCIENCE AND TECHNOLOGY

GOAT MILK CHEMISTRY AND ITS PRODUCT MANUFACTURING TECHNOLOGY

FOOD SCIENCE AND TECHNOLOGY, MANUFACTURING TECHNOLOGY RESEARCH

Additional books and e-books in this series can be found on Nova's website under the Series tab.

MANUFACTURING TECHNOLOGY RESEARCH

Additional books and e-books in this series can be found on Nova's website under the Series tab.

FOOD SCIENCE AND TECHNOLOGY

GOAT MILK CHEMISTRY AND ITS PRODUCT MANUFACTURING TECHNOLOGY

MINGRUO GUO
EDITOR

nova
science publishers
New York

Library of Congress Cataloging-in-Publication Data

Names: Guo, Mingruo, 1960- editor.
Title: Goat milk chemistry and its product manufacturing technology / Mingruo Guo, editor.
Description: New York : Nova Science Publishers, [2019] | Series: Food science and technology | Includes bibliographical references and index. |
Identifiers: LCCN 2019060044 (print) | LCCN 2019060045 (ebook) | ISBN 9781536172997 (hardcover) | ISBN 9781536173000 (adobe pdf)
Subjects: LCSH: Goat milk. | Goat milk industry. | Goats--Nutrition. | Dairy products--Quality control. | Dairying.
Classification: LCC SF388 .G63 2019 (print) | LCC SF388 (ebook) | DDC 637/.17--dc23
LC record available at https://lccn.loc.gov/2019060044
LC ebook record available at https://lccn.loc.gov/2019060045

Published by Nova Science Publishers, Inc. † New York

CONTENTS

PREFACE

Goats are one of the most adaptable farm animals in the world since they can survive almost every corner of the planet. There are about 210 million dairy goats in the world. More people drinking goat milk than the milk of any other farm animals on this planet. However, the investment resources on dairy goat industry and research on the chemical, nutritional, and biological properties of goat milk and its product developments are still very limited. Although books related to goat milk and goat milk processing technology are available on the market, but there is still no such book in English specifically dealing with goat milk chemistry and its products manufacturing technology.

I have been teaching and doing research on goat milk chemistry and goat milk products processing for more than 30 years both in China and the United States. I decided to take one half Sabbatical leave to write/edit a book on this topic: Goat Milk Chemistry and its Products Manufacturing Technology.

The content of this book includes eight chapters: Introduction; Goat milk production and consumption; Goat milk chemistry and physical properties; Goat milk flavor; Goat milk products and manufacturing technology; Goat milk product quality control; Goat milk nutrition and health benefits; and Outlook of goat milk production, products and market.

This book is the results of my collaboration with 15 other contributors from academia, the food industry, and my former and current Ph.D. students in this field. I sincerely thank the contributors for their hard work, dedication, and willingness to collaborate with me on this project. It is my hope and intention that this book would provide the readers with a set of comprehensive and updated information about goat milk chemistry and goat milk products manufacturing technology.

In: Goat Milk Chemistry …
Editor: Mingruo Guo
ISBN: 978-1-53617-299-7

Chapter 1

INTRODUCTION AND BACKGROUND

Mingruo Guo*
Department of Nutrition and Food Sciences,
College of Agriculture and Life Sciences,
University of Vermont, Burlington, Vermont, US
Key Laboratory of Dairy Science, Northeast Agricultural University,
Ministry of Education, Harbin, Heilongjiang, China

According to Merriam-Webster Dictionary, goat is any of various hollow-horned ruminant mammals related to sheep that have backward-curving horns, a short tail and usually straight hairs. Goats might be among the first domesticated animals by mankind back to 10,000 years ago in Iran, Iraq, Syria, and Eastern Turkey (Zeuner 1963). Goats are one of the most adaptable farm animals in the world since they can survive in almost every corner of the planet.

Goats are considered as the "poor man's cow". Although most goats are in developing countries, breeding and new developments are mainly from Europe and North America.

* Corresponding Author's Email: mguo@uvm.edu.

Table 1. Selected dairy goat breeds and their average milk production

Breed	Milk yield (kg/year)
Saanen	1200
Alpine	1140
Toggenburg	1030
LaMancha	1000
Nubian	900

Adapted from Starbard (2015).

During the last few centuries, dairy goat breeds have been specifically developed around the world. Milk production efficiency has been much improved using genetic selection and modern feeding management technology (Haenlein 2007). The major dairy goats and their average annual milk yield are shown in Table 1.

There are about 210 million dairy goats in the world. Asia and Africa are the main goat milk producing regions that accounts for almost 80% of total population of dairy goats. Europe and Americas have about 5% and 4%, respectively (FAO 2019). Indian subcontinent has nearly 37% of the world's milk goats and produced more than 40% of the total milk for the world. China is raising around 1.5 million dairy goats and produce nearly 2% of total goat milk production.

The total goat milk production in Africa was 3.9 million tons and there were about 80 million milk goats in 2017 (FAO 2019). Europe produces about 17% of the world's goat milk with only 4% of diary goats due to the fact that its milk yield is 4 times higher than the average of the world. France, Spain, Greece and Turkey are the leading countries for goat milk production. According to 2017 Census of Agriculture by the USDA (2019), there are about 373,000 dairy goats in the United States of America and milk goats are found in all the 50 states. Wisconsin has the most with about 44,000 and California has about 41,000. Iowa is the third and has 30,000 milk goats. The dairy goat farming sector grew faster than any other major livestock farming between 2007 and 2017.

Table 2. Distribution of dairy goats of the five continents (%)

Continent	Distribution (% of total)
Asia	52
Africa	39
Europe	5
Americas	4
Oceania	1

Adapted from Miller and Lu (2019).

The global dairy goat population has continued to grow during past three decades, but the overall distribution of dairy goats has not changed much among the major continents. The distribution of milk goats in the world was shown in Table 2. Although goat milk production ranks the third after bovine and buffalo milk, there are more people drinking goat milk than the milk of any other farm animals on this planet (Guo 2003). Goat milk production and consumption play an important role in the diet and economic aspects for many developing countries. Goat milk and its products provide valuable and high quality protein, minerals, and other nutrients for the people in Middle East, South Asian countries, and certain tropical regions. Goat milk is often used as an appropriate substitute for bovine milk in situations where the latter may cause allergic responses to some consumers. Modern dairy processing technology makes it possible to manufacture all the dairy products the same as cow milk. Goat milk products are considered as specialty dairy foods and are becoming more popular in North America and other developed countries. Goat milk will continue to grow across the globe in the years to come.

REFERENCES

Food and Agriculture Organization of United Nations (FAO) (2019). *Food and Agriculture Organization of United Nations Statistical Databases*. Http://Fao.Org.

Guo, M. R. (2003). *Goat Milk. Encyclopedia of Food Sciences and Nutrition,* 2nd Edition, Caballero, B. (Ed.), Volume 5. Pp. 2944 - 2949. Academic Press, New York, NY, U.S.A.

Haenlein, G. F. W. (2007). About The Evolution of Goat and Sheep Milk Production. *Small Ruminant Research*, 68(1 - 2), 3 - 6.

Miller, B. A. and Lu, C. D. (2019). Current Status of Global Dairy Goat Production: An Overview. *Asian-Australian Journal of Animal Sciences,* 32 (8):1219 - 1232. Special Issue.

Starbard, Ann (2015). *The Dairy Goat Handbook.* Quarto Publishing Group USA Inc. Minneapolis, MN, USA, P30.

USDA (United States Department of Agriculture). (2019). National Agricultural Statistics Service: 2017 Census of Agriculture.

Zeuner, F. E. (1963). *A History of Domesticated Animals*. Harper & Row Publishers, New York, P560.

In: Goat Milk Chemistry …
Editor: Mingruo Guo
ISBN: 978-1-53617-299-7

Chapter 2

GOAT MILK PRODUCTION AND CONSUMPTION

Hengbo Shi[1,2], Huaiping Shi[3], Jun Luo[3], Fengqi Zhao[1,4] and Mingruo Guo[5,6]*

[1]College of Animal Sciences, Zhejiang University, Hangzhou, Zhejiang, China
[2]College of Life Sciences, Zhejiang Sci-Tech University, Hangzhou, Zhejiang, China
[3]College of Animal Science and Technology, Northwest A&F University, Yangling, Shanxi, China
[4]Department of Animal and Veterinary Sciences, University of Vermont, Burlington, VT, US
[5]Key Laboratory of Dairy Science (Northeast Agricultural University), Ministry of Education, Harbin, Heilongjiang, China
[6]Department of Nutrition and Food Sciences, College of Agriculture and Life Sciences, University of Vermont, Burlington, Vermont, US

* Corresponding Author's Email: mguo@uvm.edu.cn.

ABSTRACT

Dairy goats are widely reared worldwide and are often distributed throughout low-income, food-deficit countries as a primary human food resource. Due to its nutritional value and market demand, goat milk and its products play an important role in human health and economic viability in many parts of the world. In this chapter, we provide an overview of the global distribution of dairy goats, goat lactation physiology and the economy of dairy goat industry in the major dairy goat-farming countries. The total number of dairy goats in the world is approximately 203 million. Most dairy goats are raised in Asia and Africa. Sixty-one percent of the dairy goats worldwide are raised in tropical and arid regions. Due to their production performance, the five major dairy goat breeds (Alpine, Saanen, Toggenburg, LaMancha, and Nubian) recognized by the American Dairy Goat Association have been crossbred with local goats to form various new dairy goat breeds that are distributed globally. Similar to cow milk, protein, lactose and fat are the three major nutritional components in goat milk; however, goat milk has some differences in its protein and fatty acid profiles, making it a preferable food source for people who are allergic to cow milk. Goat milk and its products play an important role in economic viability in many parts of the world, especially in developing countries.

Keywords: dairy goat, goat milk, milk composition, goat milk economics

INTRODUCTION

Goat was the first domesticated ruminant. Due to its tolerance to poor food supply, small body size and easy management, dairy goats are bred globally to provide humans with milk, much like dairy cows. However, dairy goats have distinct advantages in areas with harsh climate conditions and steep mountains or high altitudes, such as in France or Sudan. Because of their easy management and tolerance to poor food sources, dairy goats are more widely distributed throughout low-income, food-deficit countries, where their products are a primary food source. However, dairy goats are also present in high-income and developed countries, including France, Germany, the US and Canada.

Goat milk and its products have played an important role in economic viability in many parts of the world. Compared with bovine milk, milk from small ruminants is mainly used to produce niche or ethnic dairy foods, which are usually used for self-consumption or sold as traditional and high-quality dairy products (e.g., fresh cheeses and yogurt) (Dennett, 2016). Pasteurized goat milk has become popular in Asia and South America in recent years.

Given the growing importance of the global dairy goat industry, this chapter aims to discuss current dairy goat distribution, milk production and dairy goat economics worldwide.

DAIRY GOAT DISTRIBUTION

Table 1. Distribution of the goats in the world[1]

Continent	Total [million head (%)][2]	Dairy [million head (%)]	Milk [Mt (%)]	Yield [L/head]
Asia	556 (55.4)	106 (52.1)	8.04 (52.7)	76.2
Africa	388 (38.7)	80.4 (39.6)	3.93 (25.7)	48.9
Europe	17 (1.7)	9 (4.3)	2.54 (16.6)	290.1
America	38 (3.8)	8 (4.0)	0.75 (4.9)	93.4
Oceania	4 (0.4)	<0.1 (0)	<0.01 (0)	-
Total	1, 003 (100)	203 (100)	15.26 (100)	75.3[3]

[1] Source: FAOSTAT (2018). This table is modified from a paper by (Pulina et al., 2018).
[2] Total number of goats (regardless of purpose).

Dairy goat farms are mainly distributed in subtropical-temperate areas of Asia, Europe, and Africa, most often in low-income, food-deficit countries. According to FAOSTAT (Food and Agriculture Organization of the United Nations Statistics database, http://www.fao.org/faostat/en/#data), the total number of goats in the world is approximately one billion in 2018, of which dairy goats compose approximately 20% (203 million). Most dairy goats (92%) are raised in Asia (106 million, 52.1%) and Africa (80.4 million, 39.6%), with smaller populations in Europe (9 million 4.3%), the Americas (8 million, 4.0%), and Oceania (< 0.1%) (Pulina et al., 2018)

(Table 1). The tropical and arid regions of the world contain 61% of the world's goats.

Dairy Goat Breeds and Their Milk Production Capacity

The capacity of dairy goats to provide milk is determined by their breed. Each breed also has a reputation associated with the flavor of its milk, which, in combination with its butterfat content, may determine the breed dynamics of a working dairy herd. Currently, there are at least 80 breeds of dairy goats in the world.

Because of the emphasis on the breed standards across regions in the dairy goat sector, there is likely little incentive to import goats or introduce new genes from other countries or regions. The American Dairy Goat Association (ADGA) recognizes nine dairy goat breeds, including Alpine, Saanen, Toggenburg, LaMancha, Nubian, Oberhasli, Dwarf Goat, Nigerian Dwarf, and Pygmy. The following five major global breeds and their milking characteristics are discussed.

Alpine

The Alpine, which originated in the French Alps, is a breed of domestic goat with excellent milking ability. It is a medium- to large-sized breed with a variety of coat colors and upright ears (Figure 1). Alpines can produce milk over an extended period of time and are thus widely used globally as tupping rams for crossbreeding to improve the production of local dairy goats. The Alpine is the second most registered dairy goat.

The average yearly milk production per doe is 1017 kg, with an average 3.4% milk fat content according to the National Dairy Herd Improvement statistics (Table 2). The highest milk production record for the breed is over 2300 kg a year.

Figure 1. The Alpine goat (adapted from the American Dairy Goat Association, http://adga.org/breed-standards/).

Table 2. Milk production by major dairy goat breed (DHIR data)

Breed	Average production (kg/lactation)	Production range (kg/lactation)	Milk fat (%)	Milk protein (%)
Alpine	1017	356-2462	3.4	2.9
LaMancha	945	333-1944	4.0	3.2
Nubian	819	252-1922	4.9	3.7
Saanen	1160	275-2471	3.3	2.9
Toggenburg	952	423-1971	3.2	2.7

DHIR= Dairy Herd Improvement Registry (http://adga.org/).

Saanen

The Saanen originated in the historic region of the Saanenland in western Switzerland. They are either white or cream-colored and also have upright ears (Figure 2). They are productive milkers, second only to Alpines. Because of their high productivity, Saanen goats have been reared in many countries worldwide since the nineteenth century.

Many local sub-breeds are the result of crossbreeding of Saanens with local goats. Among these local variants are the Banat White in Romania, British Saanen, French Saanen, Israeli Saanen, Russian White, Laoshan and Guanzhong Dairy Goats in China, and Yugoslav Saanen. Until 2016, Saanen goat was found in more than eighty countries. The total world population of Saanen goats is reported to be over 900,000 (http://www.fao.org/dad-is/zh/). Saanen milk has a butterfat content of 3.3% with an annual average production of 1160 kg per doe (Table 2).

Figure 2. The Saanen goat (adapted from the American Dairy Goat Association, http://adga.org/breed-standards/).

LaMancha

The LaMancha goat is the only breed of dairy goat developed in the United States and is also known as American LaMancha. Coats of any color are acceptable, but LaMancha are distinctive because they lack external ears (Figure 3).

The breed standard is very specific regarding the characteristics of the LaMancha ear, with only the gopher type being eligible for buck registration. LaManchas are also known for their high milk production, relatively high butterfat content in their milk and a greater than 10-month lactation period. Their average yearly milk production is 945 kg per doe, with a butterfat concentration of 4.0% (Table 2).

Figure 3. The LaMancha goat (adapted from the American Dairy Goat Association, http://adga.org/breed-standards/).

Nubian

Nubian goats are probably native to Africa but are also common in India and the Middle East and have been since ancient times. Nubians have a large body size, can be of any color, have long ears that are pendulous and floppy, and have a very round nose (referred to as a 'Roman Nose') (Figure 4). Due to their heritage, Nubians can live in very hot climates and have a longer breeding season than other dairy goats. Nubian milk has an average of 5% or more butterfat and is thus commonly used in large-scale milk production, such as for cheese making. The average milk production of Nubian goats is approximately 820 kg per doe over a 6-month lactation period (Table 2).

Figure 4. The Nubian goat (adapted from the American Dairy Goat Association, http://adga.org/breed-standards/).

Toggenburg

The Toggenburg goat is the oldest known dairy breed and originated in the Toggenburg valley in Switzerland. They are medium-sized goats with does weighing at least 50 kg. Toggenburg goats are brown (fawn to dark chocolate) with distinctive white markings on the face, ears, muzzle, legs and tail (Figure 5). The Toggenburg is among the most productive breeds of dairy goat and is distributed in 50 counties worldwide. These goats produced an average of 952 kg milk per doe with a butterfat content of 3.2% (Table 2).

Figure 5. The Toggenburg goat (adapted from the American Dairy Goat Association, http://adga.org/breed-standards/).

All of the above five breeds are distributed worldwide. Because of their high production and easy management, they have been crossbred with local goats to form various new dairy goat breeds.

Overview of Dairy Goat Production Worldwide

Goat milk production is a dynamic and growing industry. In many developing countries, goat milk is the primary milk source for humans. However, compared with milk from dairy cows, goat milk is generally not consumed directly as liquid milk but is commonly used in the production of niche or ethnic dairy foods, e.g., cheese and yogurt (Dennett, 2016). Nevertheless, with a change in diet structures, liquid goat milk is increasingly consumed by humans as a supplementary milk or food source for its health benefits or, in some developed countries, for its unique taste.

Data from the FAOSTAT (2018) showed that in 2018, the majority of goat milk (total 15.3 Mt) was produced in Asia (8.04 Mt, 52.7%), with significant amounts in the Indian subcontinent (i.e., India, Bangladesh, and Pakistan) and China, followed by Africa (3.93 Mt, 25.7%), where Sudan and South Sudan were the most important milk producers. Europe (2.54 Mt, 16.6%) and America (0.75 Mt, 4.9%) contributed to the world's total goat milk production, whereas production in Oceania was negligible (< 0.1%). Over the past 50 years, worldwide goat milk production has more than doubled worldwide, even though there has been a small decrease since 2016 (Figure 6). If the increasing trend of the last 50 years is maintained, it is expected that milk production will increase by approximately 9.7 Mt (+53%) by 2030 (Pulina et al., 2018).

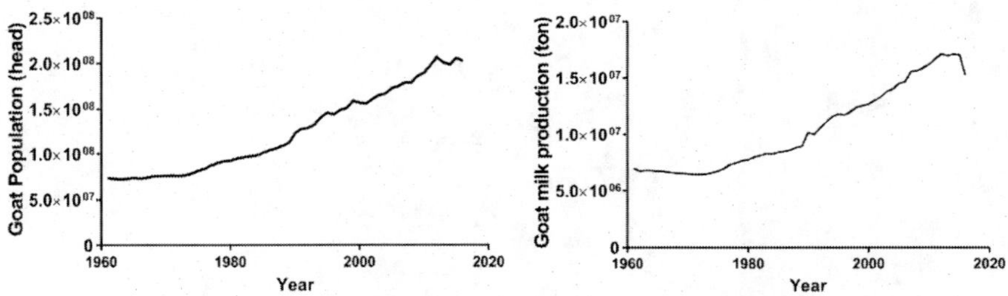

Figure 6. Global changes in goat population and milk production from 1960 to 2016 (Data from FAOSTAT, 2018).

Although the population record of each breed is limited, the six major breeds of dairy goats and their crossbreeds contribute to the total production of goat milk of the world. In terms of milk fat content, Nubian goat milk has the highest concentration of milk fat (4.9%) and protein (3.7%) (Table 2).

According to FAOSTAT (2018), the average milk yield of dairy goats varies widely by continent (49 to 290 L/year), depending on goat breeds and farm management practices. Countries of the Indian subcontinent rear 36.7% of the world's dairy goats and produce 40.7% of the world's goat milk, with the highest production in India (60.6%; 129 L/year), Bangladesh (16.9%; 37

L/year), and Pakistan (13.3%; 100 L/year). The number of dairy goats in India and Bangladesh increased by more than three times during the last 50 years and reached 29 and 1 million heads, respectively (Figure 7). India has an obvious advantage in the dairy goat number among the Indian subcontinent countries. Along with the goat population increase, the goat milk production in India and Bangladesh increased by more than 4 times during the last 50 years with the increase of milk yield. However, compared with Europe, milk yield in Indian subcontinent countries is still low and has substantial room for improvement.

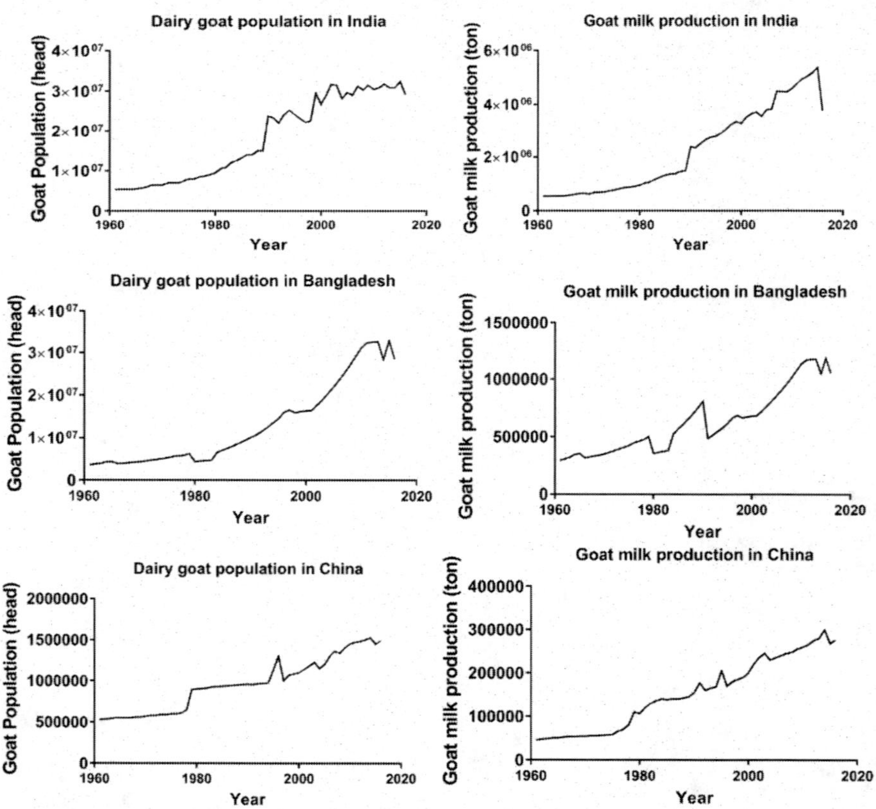

Figure 7. Goat populations and goat milk production in India, Bangladesh and China from 1960 to 2016 (Data from FAOSTAT, 2018).

China has approximately 1.5 million dairy goats that produce nearly 2% of the world's goat milk with a moderate average milk yield of 189 L/year. In the last 50 years, the dairy goat population and total goat milk production in China increased approximately 3- and 4-fold, respectively (Figure 7). However, compared with the Indian subcontinent countries, the population of dairy goats is still low in China. Farming of dairy goats is regionalized, mainly concentrated in arid and semiarid areas where the economy is less developed. Dairy goats are raised by farmers to supplement their income through milk sales. Although large-scale dairy goat farms (>2,000 head) have been established in some areas, such as Shanxi Province, in the past five years, small family farms account for more than 85% of the total milk production. In recent years, some Chinese provincial governments have vigorously supported the development of the dairy goat industry. However, considering that the feeding and breeding of dairy goats are mainly conducted on family farms, the technological improvement of dairy goat production still has a long way to go to improve milk yields.

In Africa, the total goat milk production was 3.9 million tons, and the population of dairy goats was 80.3 million according to FAOSTAT (2018). The major countries for dairy goat farming in Africa are Sudan and Southern Sudan, which had 1.1 million and 466,000 heads of dairy goats in 2016, respectively. FARM-Africa, an active member of the International Goat Association and who founded the East Africa Goat Development Network (EAGODEN), contributed greatly to the development of dairy goat production in rural Africa. FARM-Africa promotes education and best practices across the region to reduce poverty and improve the nutrition and health of women and children at the household level through dairy goat rearing (Peacock, 2008). Currently, there is a demand for 'improved' dairy goat breeds in Africa. European dairy goat breeds have been imported to crossbreed with local ones to improve the production of dairy goats. These projects will help the development of the dairy goat industry in the future. According to FAOSTAT (2018), dairy goat populations in Africa increased by approximately four times in the past 50 years, and goat milk production had a similar increasing pattern (Figure 8). It is predicted that the increase in dairy goat production will contribute to the reduction of poverty in Africa (Bett, Kosgey, Kahi, & Peters, 2009).

According to FAOSTAT (2018), the dairy goat population and milk production in the Americas has increased over the past 50 years (Figure 8). Compared with Asia and Africa, the trend was slower. In 2016, milk production in the Americas reached approximately 0.75 million tons. Most goat milk was used to make cheese for western food production.

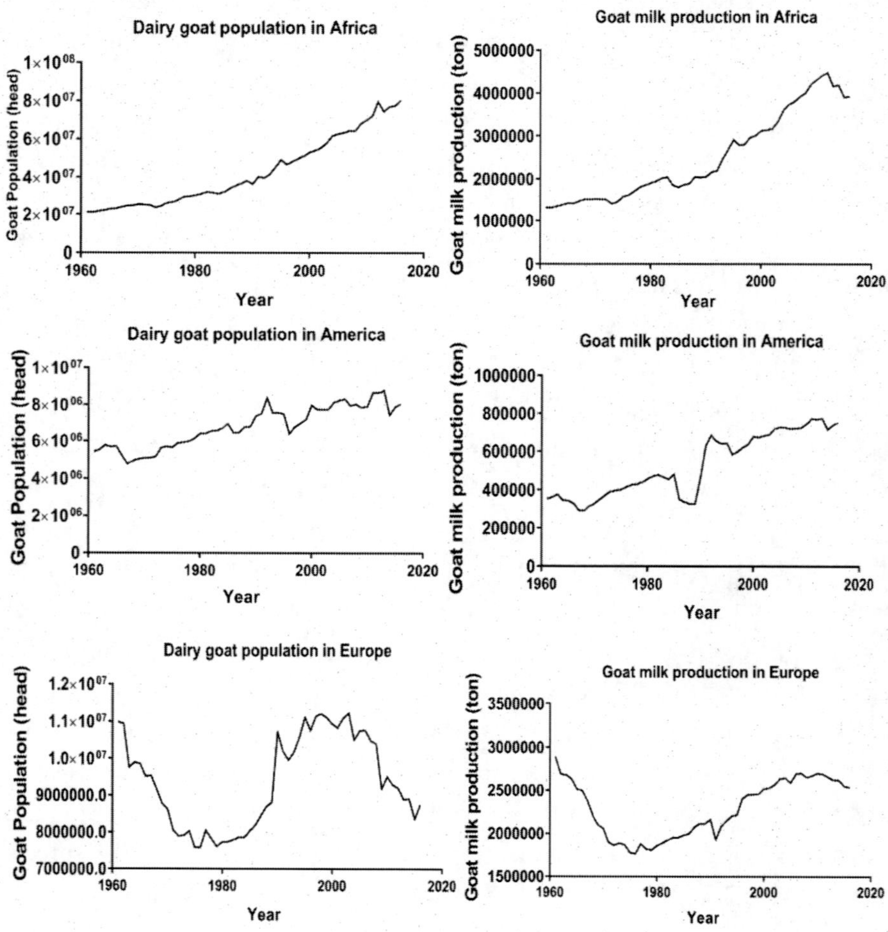

Figure 8. Changes in dairy goat populations and goat milk production of Africa, the America and Europe from 1960 to 2016 (Data from FAOSTAT (2018).

Dairy goat breeding originated in Europe. Europe produces 16.6% of the world's goat milk with only 4.3% of the dairy goat population (Table 1, Figure 8). The milk yield per doe is almost 4 times the average yield worldwide. Countries in the Mediterranean and Black Sea regions, such as France (20.7%; 715 L/year), Spain (14.1%; 328 L/year), Greece (13.2%; 134 L/year), and Turkey (11.8%; 73 L/year), are the current leaders in goat milk production, followed by the Russian Federation (8.6%; 330 L/ year), Ukraine (7.8%; 500 L/year), and Algeria (7.9%; 85 L/year) (International Committee for Animal Recording, ICAR, 2018, https://www.icar.org/). In Europe, although most goat milk is used for cheese production, its consumption as whole milk and yogurt is also common. Even though specific data on the international goat cheese trade are not available, more than 50% of protected denomination of origin goat cheeses are produced in Spain or France. Goat milk is also directly consumed by people intolerant to cow milk or suffering from digestive disorders in Spain.

LACTATION PHYSIOLOGY IN DAIRY GOATS

Table 3. Average chemical compositions and energy of goat and cow milk

	Cow	Goat
Energy (kcal/100 ml)	69.00	70.00
Lactose (%)	4.70	4.10
Protein (%)	3.50	3.20
Fat (%)	3.60	3.80
Cholesterol (mg/100 ml)	15.00	12.00
Calcium (%)	0.18	0.19
Iron (%)	0.06	0.07
Vitamin A (IU/g fat)	21.00	39.00
Vitamin B (IU/g fat)	0.70	0.70
Vitamin C (IU/g fat)	2.00	2.00
Riboflavin (μg/100 ml)	159.00	210.00

Data from http://www.ADGA.org/

Similar to cow milk, the nutritional aspects of goat milk are derived mainly from proteins, fats, lactose, minerals and vitamins. As shown in Table 3, goat milk is generally similar in composition to cow milk but contains less lactose, more calcium and potassium, and more vitamin B6 and vitamin A than cow milk.

In addition, the difference in lactation physiology between the two ruminant species results in some important differences in the protein structure and fatty acid profile. Because of these differences, people who have allergies to cow milk may be able to drink goat milk, and the small size of fat globules in goat milk stay in suspension longer, which leads to the perception of "natural homogenization" and the "easy digestion" of goat milk.

Milk Fat

Among mammalian species, the lactating mammary gland is the most lipogenic organ of the body. Lipids in milk supply the majority of the calories required for neonatal growth in many species (Oftedal, 2000) and are a primary source of the essential fatty acids for neonatal membrane synthesis and synthesis of eicosanoids and other bioactive lipid signaling molecules.

In ruminants, approximately half of the milk fat (mainly short- and medium-chain fatty acids) is synthesized de novo by the mammary gland, and the other half (mainly long-chain fatty acids) is directly taken from the blood (Harvatine, Boisclair, & Bauman, 2009). For the synthesis of lipids in goat milk, a complex gene network related to lipid metabolism in goat mammary glands has been described (Figure 9) (Shi et al., 2015). In this gene network, two members of the fatty acid binding protein (FABP) family, FABP3 and FABP4, are highly expressed in the lactating goat mammary gland. The biological function of FABPs is to transport fatty acids to the sites of triacylglycerol or phospholipid synthesis (Calvo, Marcos, Jurado, & Serrano, 2004; McArthur et al., 1999). The high expression of FABPs indicates their essentiality for the uptake and trafficking of fatty acids in the mammary gland.

The enzyme fatty acid synthase (gene: FASN), a rate-limiting enzyme for fatty acid de novo biosynthesis during lactation (Zhu et al., 2014), also has high expression in the mammary gland. The high expression of FASN could partly explain why goat milk is rich in short- and medium-chain fatty acids (Morris et al., 2007; Silanikove, Leitner, Merin, & Prosser, 2010). SCD (Δ-9-desaturase), an endoplasmic reticulum-bound enzyme, catalyzes the desaturation of saturated fatty acyl-CoAs (Yao et al., 2017), which can directly affect the composition of unsaturated fatty acids in milk. During lactation, fatty acids are esterified into triacylglycerol within mammary cells.

Genes related to triacylglycerol biosynthesis, including glycerol-3-phosphate acyltransferase 1 (GPAT1), glycerol-3-phosphate acyltransferase 4 (GPAT4) and 1-acyl-sn-glycerol-3-phosphate acyltran-sferase alpha (AGPAT1), are also among the genes highly expressed in the mammary gland. Finally, triacylglycerol is packaged into a lipid droplet and then secreted out of the mammary cells (Shi et al., 2015).

Lactose

Lactose, commonly referred to as milk sugar, is the major carbohydrate in milk. It is synthesized by lactose synthase in the Golgi apparatus of mammary epithelial cells using glucose as the major substrate. Goat milk has less lactose than cow milk (Jelert, 1984), which may be helpful for people with intestinal disorders or lactose intolerance (Pieniak-Lendzion & Niedziolka, 2004).

In general, the expression of genes related to lactose synthesis is much lower than those related to lipid synthesis, except for LALBA (alpha-lactalbumin). LALBA, one of two subunits of lactose synthase and a classical marker of lactation, is highly expressed in goat (Shi et al., 2015), bovine (Wickramasinghe, Rincon, Islas-Trejo, & Medrano, 2012) and human (Lemay, Ballard, et al., 2013) mammary glands.

Milk Protein

In ruminant milk, more than 95% of the proteins are encoded by 6 genes: CSN1S1, CSN2, CSN1S2, CSN3, LALBA and LBG. The expression levels of these genes are high, and their transcripts often account for 70–80% of total mRNA transcripts in the lactating cow mammary gland (Lemay, Hovey, et al., 2013). These six genes share similar expression patterns in the different physiological stages (Shi et al., 2015). However, recent data of the goat mammary gland showed that the average expression of these six genes during the whole lactation period in goats accounts for 42% of all mRNA transcripts (Shi et al., 2015), which is lower than previously reported in lactating cow mammary glands (Lemay, Hovey, et al., 2013).

Milk protein αs1-casein is associated with milk allergy. The level of αs1-casein in goat milk is associated with polymorphisms within the CSN1S1 gene (Montalbano, Segreto, Di Gerlando, Mastrangelo, & Sardina, 2016). At least 18 alleles have been detected thus far and have been shown to be associated with four different levels of αs1-casein, ranging from 0 (e.g., CSN1S1 0) to 3.6 g/l per allele (e.g., CSN1S1 A and B), in goat milk (Ollier, Chauvet, Martin, Chilliard, & Leroux, 2008). The mRNA expression of CSN1S1 is high in goat mammary gland, but the CSN1S1 polymorphism has a major effect on the amount of milk proteins and lipids and thus affects the processing and sensory properties of milk (Ollier et al., 2008). It has been shown that four casein transcripts (CSN1S1, CSN1S2, CSN2, and CSN3) have similar levels of abundance in goat, sheep and cow mammary tissues (Bevilacqua, Helbling, Miranda, & Martin, 2006; Colitti & Pulina, 2010; Sigl, Meyer, & Wiedemann, 2012), but the amounts of the corresponding proteins are very different, suggesting that the four casein mRNAs are translated with different efficiencies (Bevilacqua et al., 2006). The high expression of caseins and whey protein genes at 310 days of parturition supports previous observations that casein-encoding genes are highly expressed in goat mammary glands at every stage of lactation (Bevilacqua et al., 2006). The mechanism behind the different translation efficiencies of milk genes is unknown.

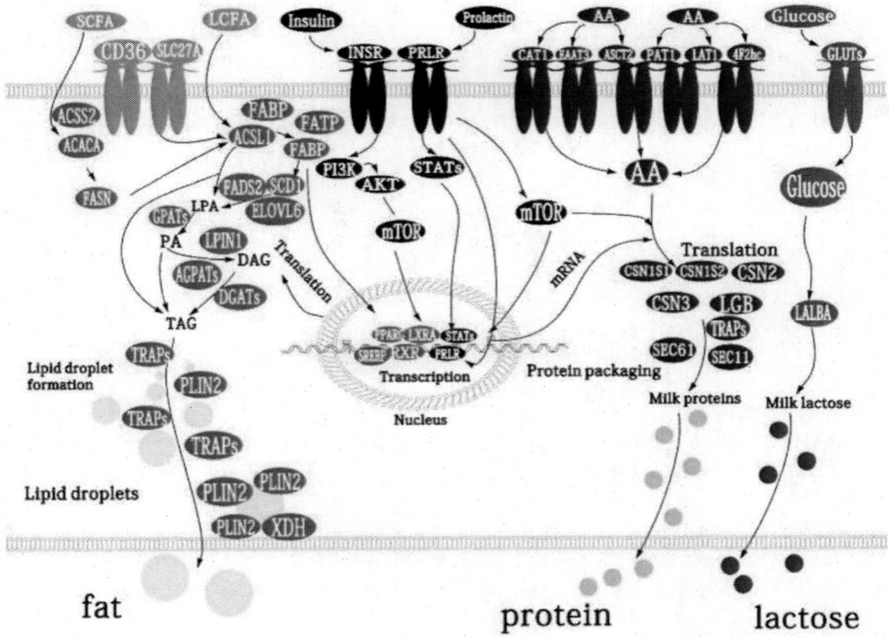

Figure 9. A model to illustrate the key factors potentially involved in the regulation of milk fat, protein and lactose biosynthesis in goat mammary glands. The model includes networks involved in the uptake of fatty acids, amino acids and glucose; signaling pathways of insulin and prolactin; and biosynthesis and secretion processes of fat, protein and lactose (adapted from Shi et al., 2015). The gene abbreviations are listed at the end of the chapter.

Gene abbreviations in Figure 9:

CD36	CD36 molecule (thrombospondin receptor)
ELOVL6	Elongation of very-long-chain fatty acids protein 6
FABP3	Fatty acid-binding protein 3
FABP4	Fatty acid-binding protein 4
ACSS2	Acetyl-coenzyme A synthetase, cytoplasmic-like isoform 2
ACSS1	Acetyl-coenzyme A synthetase, cytoplasmic-like isoform 1
ACSL1	Long-chain fatty acid CoA ligase 1
ACACA	Acetyl-CoA acetyltransferase
FADS1	Fatty acid desaturase 1
FADS2	Fatty acid desaturase 2
FASN	Fatty acid synthase
SCD	Stearoyl-CoA desaturase
SLC27A1	Long-chain fatty acid transport protein 1
PLIN1	Perilipin 1
PLIN2	Perilipin 2

(Continued).

Gene abbreviations in Figure 9 (Continued):

XDH	Xanthine dehydrogenase
AGPAT1	1-acyl-sn-glycerol-3-phosphate acyltransferase alpha
AGPAT3	1-acyl-sn-glycerol-3-phosphate acyltransferase gamma
AGPAT4	1-acyl-sn-glycerol-3-phosphate acyltransferase delta
AGPAT6	Glycerol-3-phosphate acyltransferase 4
DGAT1	Diacylglycerol O-acyltransferase 1
DGAT2	Diacylglycerol O-acyltransferase 2
GPAM1	Glycerol-3-phosphate acyltransferase 1
B4GALT1	Beta-1,4-galactosyltransferase I
UGP	UTP--glucose-1-phosphate uridylyltransferase
INSR	Insulin receptor
LXRA	Liver X receptor alpha
PPARG	Peroxisome proliferator-activated receptor gamma
PRLR	Prolactin receptor
RXR	Retinoic acid receptor RXR-beta
SREBP1	Sterol regulatory element-binding protein 1
STAT5a	Signal transducer and activator of transcription 5A
STAT5b	Signal transducer and activator of transcription 5B
CAT1	Carnitine O-acetyltransferase
EAAT3	Solute carrier family 1
CSN2	Beta-casein
CSN3	Kappa-casein
LGB	Beta-lactoglobulin
LALBA	Alpha-lactalbumin
CSN1S1	Alpha-S1-casein
CSN1S2	Alpha-S2-casein
SEC11C	Signal peptidase complex catalytic subunit
Sec61	Protein transport protein Sec61 subunit alpha isoform 1
TRAPb	Translocon-associated protein subunit beta isoform 1
TRAPd	Translocon-associated protein subunit delta
TRAPg	Translocon-associated protein subunit gamma
B4GALT1	Beta-1,4-galactosyltransferase I
GLUT1	Solute carrier family 2, facilitated glucose transporter member 1
GLUT4	Solute carrier family 2, facilitated glucose transporter member 4
UGP	UTP--glucose-1-phosphate uridylyltransferase

Many milk components, such as protein, lactose and citrate, are secreted via vesicle trafficking (Lu, van Hooijdonk, Boeren, Vervoort, & Hettinga, 2014; McManaman & Neville, 2003). In mammalian cells, secretory proteins are typically translocated across the endoplasmic reticulum (ER) membrane in a co-translational mode by the ER protein translocon, which comprises the protein-conducting channel Sec61 and translocon-associated protein complex (Nagasawa, Higashi, Hosokawa, Kaufman, & Nagata, 2007). The mammalian translocon-associated protein (TRAP) complex comprises four transmembrane protein subunits, including α-, β-, γ- and δ-subunits, in the ER (Nagasawa et al., 2007). The milk fat fraction contains PLIN2, a protein that coats lipid droplets and is involved in milk fat globule trafficking and secretion (Chong, Reigan, Mayle-Combs, Orlicky, & McManaman, 2011; Lemay, Ballard, et al., 2013). These genes are all highly expressed to support active intracellular trafficking, transport and secretion of protein, lactose and lipid droplets in goat mammary epithelial cells (Chong et al., 2011; Shi et al., 2015). Currently, approximately 1,283 annotated unigenes are related to this process (Shi et al., 2015) (Figure 9). Such a large number of unigenes annotated for intracellular trafficking, transport and secretion classification underscores the complexity and importance of this process in the mammary gland.

GOAT MILK ECONOMICS

Goat milk and its products have played an important role in economic viability in many parts of the world, especially in developing countries. Goat milk is rich in short- and middle-chain fatty acids and has a low level of α-s1-casein. Dairy goat farming has become a widespread economic activity, and many more farmers are engaging in this activity to support their livelihood.

Compared with cows, goats are cheaper to maintain and require less space. Dairy goats are mostly reared by low-income or middle-income earners and hence are named the ‘poor man's cow.’ However, large-scale dairy goat farms have been developed due to the expansion of the dairy goat industry during the past ten years.

Raising goats for milk may be less economical than raising dairy cows, considering that dairy cows produce much more milk per animal. However, the investment for a dairy goat farm is less expensive. Dairy goat farms require much less space, and even rural backyards are suitable. The average dairy cow produces 65 pounds of milk per day with 50 pounds of feed, whereas the average dairy goat needs 6 pounds of feed to produce 4-7 pounds of milk per day. A study of the management of 108 US dairy goat herds showed an income over feed cost of $117.00 to $406.00 per goat per year with the milk yield of 503 to 1,115 kg per goat per year (USDA–DHIA 2018). The break-even producer price for goat milk were determined to be $0.52 per kg milk at an annual production level of 680 kg milk and $0.26 per kg milk at an annual production level of 1,161 kg milk in the USA. In France, the net return per goat per year is $74.93 for the wholesale farm, but a farm with retail sales will have a $112.00 return when the goat milk is used for cheese products (https://countrysidenetwork.com/daily/lifestyle/home-dairy/discover-goat-milk-benefits/). However, in most countries, no break-even point has been calculated for dairy goat farms because of a lack of data and a wide diversity of production systems.

Goat milk makes up 55 percent of all milk in Bangladesh, 51 percent in Somalia, 24 percent in Iran, and 16 percent in Sudan FAOSTAT(2018), indicating the importance of goat milk and its products in developing countries with high prevalence rates of undernutrition and malnutrition. In developed countries, the demand for goat cheese products has increased.

Goat Milk Economics in Asia

In China, in the spring, many dairy goats are transported from the Northern provinces to the South, where the milk is sold at a higher price. These goats are slaughtered at the beginning of the next winter after lactation. This production model is considered to use short-sighted measures for animal husbandry. Therefore, dairy goat numbers in southern China are quite small. Goat milk production in China contributes approximately 4% of the total milk production. Goat milk is an important supplement for dairy production, particularly for residents in underdeveloped rural areas and hilly mountainous regions.

Dairy goat farming can efficiently increase a farmer's income, improve the nutrition status of the people and realize the strategic objectives of poverty alleviation. The prices of raw goat milk in China have been generally constant (¥4/kg), with the price paid to farmers by the dairy processor remaining steady during the last ten years. However, goat milk prices increased substantially in the first three months in 2018, with the procurement price nearly double in the main producing areas (up to ¥9/kg milk) due to strong demand.

In India, goat keeping is a traditional activity and a primary source of livelihood for people below the poverty line. Goat milk production is mostly confined to people who are landless and with limited access to the resources needed to rear large animals. A study in India showed that the net income per goat per year ranged from $1.80 to $10 when milk production per goat per year ranged from 178 to 595 kg, respectively. Livestock are a very real means of steady income, allowing people to pay debts and dowries and purchase farm equipment and other tools (Escareno et al., 2013).

Goat Milk Economics in Africa

In North Africa, goats provide kids, milk and fiber, especially to farmers migrating their animals through seasonal grazing lands. Goat farming in this area has not been adequately studied. There are critical factors for the integration of milk production in economic development; one of the main problems is quality assurance for manufactured products.

In West Africa, there is a great diversity of goat breeds, most of which are local breeds. Goat milk and products are served in traditional foods. Goat milk cheeses are sold at hotels and supermarkets. Different types of cheeses are prepared by tribes for local consumption.

Goats provide the main total milk supply in the East and Central African regions. The availability and distribution of milk to strategic consumers, such as low-income families, children and elderly individuals, are critical in Sudan. Goat milk cheeses in Africa are regional, as they are in many other places. For example, Sudanese braided cheese is a semihard cheese that can be prepared at home or in small dairy plants. Hard cheeses, such as Kefalotyri and Paphitico, are also produced in Sudan and take 4 to 5 months to mature. In western Kenya, a soft, fresh cheese is produced by smallholders (Escareno et al., 2013).

Goat Milk Economics in Europe

Europe is the only continent where goat milk has considerable economic importance (Escareno et al., 2013). Dairy goat breeding in Europe is most common in the Mediterranean basin. Goat production is an important economic, environmental and sociological issue for Mediterranean countries, such as Spain, France, Italy and Greece (Pulina et al., 2018).

In France, only half of French goat farms deliver their milk to dairy plants. These farms produce large amounts of goat milk and are mainly located in the areas where harvesting forages allow farmers to manage large milk production operations. In the south of France, most goat milk is used to make cheese on-farm, and less liquid milk is produced. The average price of goat milk paid to farmers by the dairy industry increased by 35% in the last decade (i.e., from €0.514/L in 2004 to €0.695/L in 2015), with an estimated break-even point of €0.85/L for farms that deliver milk to dairy enterprises and €2.96/L for on-farm cheese makers (Pulina et al., 2018). Most milk (78%) is processed into cheese or yogurt. However, on-farm cheese production is also important, accounting for 22% of French goat milk usage (Pulina et al., 2018). The consumption of UHT goat milk has increased steadily in recent years in France, reaching approximately 10 million L in 2016.

In Greece, goat milk accounted for 3.4% of the total agricultural output (TAO) and 22% of the livestock output (LO) from 2013 to 2015 (average data from ELOGAK, 2017 and ELSTAT, 2017). Only 21% (approximately 600,000 tons) of goat milk is delivered to dairy industries. The rest of the milk is used for on-farm cheese production, and the cheese is mostly sold directly to consumers. The average price of goat milk was €0.60/L in 2016 and has not changed much in Greece over the last few years. Farms are generally located in remote, mountainous and less-productive areas, where the rural economy is poorly diversified and employment opportunities are limited. Therefore, dairy goats offer rural families an opportunity to increase their income and use family labor more intensively (Theodoridis et al., 2014). Goat milk economics is gaining attention in Greece, and pasteurized goat milk and cheeses have great local importance in the current market (Pulina et al., 2018).

In Italy, goat milk accounted for only 0.2% of the TAO and 0.7% of the LO during the 2013 to 2015 period. The price of goat milk in Italy has not varied substantially over the time (Pulina et al., 2018).

In Spain, goat milk accounted for 0.7% of the TAO and 2.0% of the LO during 2013 to 2015. Among Spanish goat farms, 71% specialize in dairy production, and the remainder are dual-purpose. It is worth noting that 87.1% of goat milk is produced by one third of the farms, suggesting a high degree of dairy specialization by these farms. Goat milk prices in Spain have changed dramatically over time, increasing 27.5% from 2004 to 2016 (from €0.483/L to €0.616/L), with a peak at €0.750/L in 2015 (Dennett, 2016). Most goat milk (90.8%) in Spain is made into cheese in large dairy plants, which coexist with small local plants and artisanal farm dairies (home consumption, 7.5%; direct sales, 1.7%). Pasteurized and sterilized liquid goat milk are new options for Spanish niche consumers (e.g., organic and allergic) but are currently negligible contributors to the total production (Pulina et al., 2018).

Goat Milk Economics in the America

Although there is only a small population of dairy goats in the Americas, dairy goats play a very important role in the economy of smallholders, specifically in Latin American countries. Mexico and Brazil are the main dairy goat-producing countries according to FAOSTAT (2018). In Mexico, the dairy industry is located mainly in the regions where the majority of goat milk is produced, which has benefited farmers since large plants buy milk even from distant locations where there is a lack of services, money and technology. The dairy goat industry in this area accounts for a good portion of the economy and social activity. However, there are seasons in the year when the supply of milk saturates the industry's demand. As a result, producers experience problems in selling their milk; this has caused them to begin producing cheeses on a small scale.

In Brazil, most of the consumption of goat milk is in the form of liquid milk since only 1,100 tons (total approximately 141,000 ton) is processed annually. The most important products are milk (pasteurized, powdered milk, frozen and UHT) and fermented products (yogurts and cheeses).

The U.S. market for goat's milk is relatively small. The liquid goat milk market in the US is mainly in California, with estimated annual sales of less than $10 million (Kentucky Agricultural Development Fund [KADF], 2003).

REFERENCES

Bett, R. C., Kosgey, I. S., Kahi, A. K., & Peters, K. J. (2009). Realities in breed improvement programmes for dairy goats in East and Central Africa. *Small Ruminant Research,* 85(2-3), 157-160.

Bevilacqua, C., Helbling, J. C., Miranda, G., & Martin, P. (2006). Translational efficiency of casein transcripts in the mammary tissue of lactating ruminants. *Reproduction Nutrition Development,* 46(5), 567-578.

Calvo, J. H., Marcos, S., Jurado, J. J., & Serrano, M. (2004). Association of the heart fatty acid-binding protein (FABP3) gene with milk traits in Manchega breed sheep. *Animal Genetics,* 35(4), 347-349.

Chilliard, Y., Ferlay, A., Rouel, J. & Lamberet, G. (2003) A review of nutritional and physiological factors affecting goat milk lipid synthesis and lipolysis. *Journal of Dairy Science,* 86:1751–1770.

Chong, B. M., Reigan, P., Mayle-Combs, K. D., Orlicky, D. J., & McManaman, J. L. (2011). Determinants of adipophilin function in milk lipid formation and secretion. Trends in endocrinology and metabolism: *TEM,* 22(6), 211-217.

Colitti, M., & Pulina, G. (2010). Expression profile of caseins, estrogen and prolactin receptors in mammary glands of dairy ewes. *Italian Journal of Animal Science,* 9(3), 285-289.

Dennett, C. (2016). Key ingredients of the Mediterranean diet: The nutritious sum of delicious parts. *Today's Dietitian,* 18, 18-33.

Escareno, L., Salinas-Gonzalez, H., Wurzinger, M., Iniguez, L., Solkner, J., & Meza-Herrera, C. (2013). Dairy goat production systems: status quo, perspectives and challenges. *Tropical Animal Health Production,* 45(1), 17-34.

Harvatine, K. J., Boisclair, Y. R., & Bauman, D. E. (2009). Recent advances in the regulation of milk fat synthesis. Animal, 3(1), 40-54.

Jelert, H. (1984). Nutrition. With cow's milk allergy goat's milk is a poor alternative and ought to be discouraged. *Sygeplejersken,* 84(50), 20-22, 24.

Lemay, D., Hovey, R., Hartono, S., Hinde, K., Smilowitz, J., Ventimiglia, F., Schmidt, K., Lee, J., Islas-Trejo, A., Silva, P., Korf, I., Medrano, J., Barry, P., German, J. (2013). Sequencing the transcriptome of milk production: milk trumps mammary tissue. *BMC Genomics,* 14(1), 872.

Lemay, D. G., Ballard, O. A., Hughes, M. A., Morrow, A. L., Horseman, N. D., & Nommsen-Rivers, L. A. (2013). RNA sequencing of the human milk fat layer transcriptome reveals distinct gene expression profiles at three stages of lactation. *PLoS One,* 8(7), e67531.

Lu, J., van Hooijdonk, T., Boeren, S., Vervoort, J., & Hettinga, K. (2014). Identification of lipid synthesis and secretion proteins in bovine milk. *Journal of Dairy Research,* 81(01), 65-72.

McArthur, M. J., Atshaves, B. P., Frolov, A., Foxworth, W. D., Kier, A. B., & Schroeder, F. (1999). Cellular uptake and intracellular trafficking of long chain fatty acids. *Journal of Lipid Research,* 40(8), 1371-1383.

McManaman, J. L., & Neville, M. C. (2003). Mammary physiology and milk secretion. *Advanced Drug Delivery Reviews,* 55(5), 629-641.

Montalbano, M., Segreto, R., Di Gerlando, R., Mastrangelo, S., & Sardina, M. T. (2016). Quantitative determination of casein genetic variants in goat milk: Application in Girgentana dairy goat breed. *Food Chemistry,* 192, 760-764.

Morris, C., Cullen, N., Glass, B., Hyndman, D., Manley, T., Hickey, S., McEwan, J. C., Pitchford, W.S.,Bottema, D. K., Lee, M. H. (2007). Fatty acid synthase effects on bovine adipose fat and milk fat. *Mammalian Genome,* 18(1), 64-74.

Nagasawa, K., Higashi, T., Hosokawa, N., Kaufman, R. J., & Nagata, K. (2007). Simultaneous induction of the four subunits of the TRAP complex by ER stress accelerates ER degradation. *Embo Reports,* 8(5), 483-489.

Oftedal, O. T. (2000). Use of maternal reserves as a lactation strategy in large mammals. *Proceedings of the Nutrition Society,* 59(1), 99-106.

Ollier, S., Chauvet, S., Martin, P., Chilliard, Y., & Leroux, C. (2008). Goat's alpha(S1)-casein polymorphism affects gene expression profile of lactating mammary gland. *Animal,* 2(4), 566-573.

Peacock, C. (2008). Dairy goat development in East Africa: A replicable model for smallholders? Small Ruminant Research, 77(2), 225-238.

Pieniak-Lendzion, K., & Niedziolka, R. (2004). The importance of goat milk in human nutrition. *Wiadomosci Zootechniczne,* 42(1), 39-44.

Pulina, G., Milan, M. J., Lavin, M. P., Theodoridis, A., Morin, E., Capote, J., Thomas, D. L., Francesconi, A. H. D., Caja, G. (2018). Invited review: Current production trends, farm structures, and economics of the dairy sheep and goat sectors. *Journal Dairy Science,* 101(8), 6715-6729.

Shi, H., Zhu, J., Luo, J., Cao, W., Shi, H., Yao, D., Li, J., Sun, Y., Xu, H., Yu, K., Loor, J. (2015). Genes regulating lipid and protein metabolism are highly expressed in mammary gland of lactating dairy goats. *Functional & Integrative Genomics,* 15(3), 309-321.

Sigl, T., Meyer, H. H. D., & Wiedemann, S. (2012). Gene expression of six major milk proteins in primary bovine mammary epithelial cells isolated from milk during the first twenty weeks of lactation. *Czech Journal of Animal Science,* 57(10), 469-480.

Silanikove, N., Leitner, G., Merin, U., & Prosser, C. G. (2010). Recent advances in exploiting goat's milk: quality, safety and production aspects. *Small Ruminant Research,* 89(2-3), 110-124.

Theodoridis, A., Ragkos, A., Roustemis, D., Arsenos, G., Abas, Z., & Sinapis, E. (2014). Technical indicators of economic performance in dairy sheep farming. *Animal,* 8(1), 133-140.

Wickramasinghe, S., Rincon, G., Islas-Trejo, A., & Medrano, J. (2012). Transcriptional profiling of bovine milk using RNA sequencing. *BMC Genomics,* 13(1), 45.

Yao, D., Luo, J., He, Q., Shi, H., Li, J., Wang, H., Xu, H., Chen, Z., Yi, Y., Loor, J. J. (2017). SCD1 alters long-chain fatty acid (LCFA) composition and its expression is directly regulated by SREBP-1 and PPAR gamma 1 in dairy goat mammary cells. *Journal of Cellular Physiology,* 232(3), 635-649.

Zhu, J., Luo, J., Wang, W., Yu, K., Wang, H., Shi, H., Sun,Y., Lin, X., Li, J. (2014). Inhibition of FASN reduces the synthesis of medium-chain fatty acids in goat mammary gland. *Animal,* 8(09), 1469-1478.

In: Goat Milk Chemistry … ISBN: 978-1-53617-299-7
Editor: Mingruo Guo

Chapter 3

GOAT MILK CHEMISTRY AND PHYSICAL PROPERTIES

Ce Wang[1,2] ***and Mingruo Guo***[1,*]
[1]Department of Nutrition and Food Sciences, College of Agriculture and Life Sciences, University of Vermont, Burlington, Vermont, US
[2]Key Laboratory of Dairy Science, Northeast Agricultural University, Ministry of Education, Harbin, Heilongjiang, China

ABSTRACT

Goat milk plays an important role in human nutrition for many parts of the world. In general, there are some differences in terms of chemical composition between goat milk and cow milk. Goat milk may be a good alternative for cow's. They are more digestible and less allergic than cow milk, as the result of smaller fat globules and different content and composition in protein. And abundant vitamins and minerals are good supply for human nutrition, especially vitamin A and Ca. Furthermore, the chemistry and physical properties of goat milk can influence goat milk

* Corresponding Author's Email: mguo@uvm.edu.

product in texture and flavor and establish foundation for goat milk's future. In this chapter, chemical composition and physicochemical characteristics of goat milk will be discussed.

Keywords: protein, fat, mineral, vitamin, physicochemical characteristics

INTRODUCTION

In many developing countries, goat milk and its products play an important role in their diet and local economy. Goat milk is mainly consumed in Asia and Africa, providing rich nutrition like animal protein, calcium, and vitamin. Goat milk also can be recommended as a suitable alternative for cow milk concerning nutrition and medical value, especially when consumers are allergic to cow milk. Although the production of goat milk is increasing year by year, the marketing for goat milk still lags behind cow milk (Lad, Aparnathi, Mehta, & Velpula, 2017). Goat milk is considered by many as one of the nutraceutical drinks. Goat milk has better properties on buffering capacity, alkalinity, digestibility, and certain therapeutic values than cow milk. Nowadays, goat milk and it's dairy products are becoming more popular in many developed countries such as the USA (Guo, 2003). The goat milk industry has a bright prospect based on its nutritional value.

In this chapter, chemical composition and physicochemical characteristics of goat milk will be discussed.

GROSS COMPOSITION

Gross composition describes the chemical and physical properties on the compounds in the milk including total solids (TS), fat, crude protein, lactose and ash. The gross composition of goat milk is similar to cow milk in comparison with cow, human milk and other mammalian milks (Table 1).

The levels of TS, crude protein, and ash in goat milk are, on average, highest among three milks, especially much higher than those in human milk. The lactose content of goat milk (~4.5%) is slightly lower than cow milks (Guo, 2003). The gross compositions and yield of these milk are variable depending on internal factors such as species, breed, individuals, stage of the lactation, and condition of the udder, as well as external factors like diet, season, environment. The proportion of fat is susceptible to the above factors, ranging from 2 to 8%. Milk yield is inversely correlated with the concentrations of most gross composition in the milk like protein and fat.

As shown in Figure 1, fat and protein contents have a similar trend during the whole year. First, there was a dramatic decrease during the first 2 months, then they remained stable at the lowest concentration during the summer. After that, the contents showed an upward trend until January. (Guo et al., 2001). Goat milk and cow milk generally have three to four times the ash content of human milk, and the ash content of cow milk is closely related to the growth rate of the species designed. Therefore, infant shouldn't be fed with neither goat milk nor cow milk.

Table 1. Gross composition of goat, cow, human milk and other mammalian milk (g 100 ml^{-1})

Species	Total solids	Fat	Crude protein	Lactose	Ash
Goat	12.9	4.1	3.5	4.5	0.8
Cow	12.5	3.5	3.3	5.0	0.7
Human	12.5	4.4	1.0	6.9	0.2
Sheep	19.3	7.4	4.5	4.8	1.0
Camel	15.2	4.3	4.3	-	0.9
Buffalo	17.4	7.0	4.4	5.2	0.8
Horse	10.5	1.5	1.6	6.7	0.3
Donkey	10.8	1.8	1.7	5.9	0.4
Cat	-	10.8	10.6	3.7	1.0
Dog	22.7	9.5	7.5	3.8	1.1

Adapted from Guo (2003), Fox, McSweeney, & Paul (1998), Ahmad et al. (2008), Folin, Denis, & Minot (1919), Oftedal, Iverson, & Jensen (1995), Oftedal (1984), Oftedal & Jenness (1988).

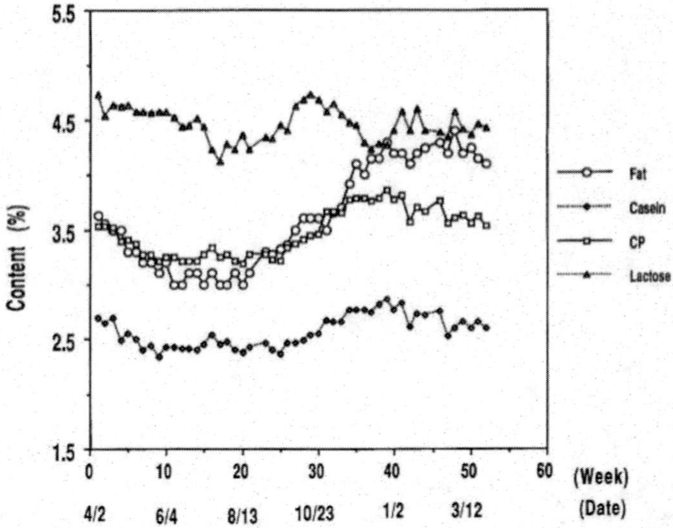

Figure 1. Changes in the contents of fat, casein, Crude Protein (CP), and lactose of goat milk during the 52 weeks (Adapted from Guo et al., (2001)).

Proteins

Nitrogen distribution in goat milk can be mainly divided into casein, serum (whey) protein and non-protein. Milk also involves important minor proteins, such as serum immunoglobulin, calcium-binding protein, lactoferrin, albumin, transferrin, prolactin and so on (Selvaggi, Laudadio, Dario, & Tufarelli, 2014). Crude protein of goat milk is in the range of 3.0% to 3.8%, higher than cow and human milk (Guo, 2003). Goat milk has a similar protein profile to cow milk, while goat milk has a different casein composition and distribution. That is the reason why goat milk has a low cheese yield and a weak yogurt structure/texture. Protein contents are influenced by species, feeding, breed, stage of lactation, season, climate and udder health status (Park, Juárez, Ramos, & Haenlein, 2007). In goat milk, caseins represent around 80% of the proteins and are classified as α_{s1}-casein (α_{s1}-CN), α_{s2}-casein (α_{s2}-CN), β-casein (β-CN), κ-casein (κ-CN), while the major whey proteins are β-lactoglobulin (β-Lg), and α-lactalbumin (α-La) and serum albumin (Table 2). Their names are given based on their homologous nature in composition and properties in corresponding proteins of cow milk (Whitney et al., 1976). Compared to cow milk, goat milk contains a higher level of the β-casein fractions, lesser levels of the αs-casein and κ-casein, especially αs1-casein.

Table 2. Protein profiles (g/100g of total protein) of goat milk and cow milk

	Goat milk	Cow milk
Total casein	81.7	82
α_{s1}-casein	16.9	32.6
α_{s2}-casein	11.5	8.5
β-casein	47.4	30.2
κ-casein	5.9	10.7
Total whey protein	18.3	18
β-lactoglobulin	10.5	10.1
α-lactalbumin	5.5	6.4
Serum albumin	2.3	1.5

Adapt from Ceballos, Sampelayo, Extremera, & Osorio (2009).

Caseins

Caseins exist in an unstable micellar state and can be precipitated from milk at specific pH whereas whey proteins remain soluble under such condition (Tsakalidou & Papadimitriou, 2016). Goat milk casein was prepared by isoelectric precipitation at pH 4.1 (Recio, Pérez-Rodríguez, Amigo, & Ramos, 1997) and cow milk casein by precipitation at pH 4.6 (Van Hekken & Thompson, 1992). Casein composition, affected by genetic polymorphism on the casein loci, is the very critical factor for casein micelle size (Tziboula-Clarke, 2003). The α-casein and β-casein are the main interior materials of the micelle, whereas κ-casein, as clusters, are located on the surface (Inglingstad et al., 2010). The diameters of casein micelles in goat milk are larger varied from 142-229 nm compared to these of cow milk, range from 142-229 nm (Ingham, Smialowska, Kirby, Wang, & Carr, 2018).

It shows a lower stability of casein micelles in goat milk, especially alcohol stability and heat stability, compared to those of cow milk (Guo, 2003). Casein micelles can diffuse light by interlinked with calcium phosphate and small amounts of potassium (K), magnesium (Mg), sodium (Na) and citrate, which results in the milk its opaque white appearance (Horáčková, Sedláčková, Sluková, & Plocková, 2014; Park et al., 2007).

Sodium dodecyl sulfate-polyacrylamide gel electrophoretograms (SDS-PAGE) shows the protein profile in goat and cow milk. The molecular weights are κ-casein, β-casein and α_s-casein from smallest to largest (Guo, 2003) (Figure 2).

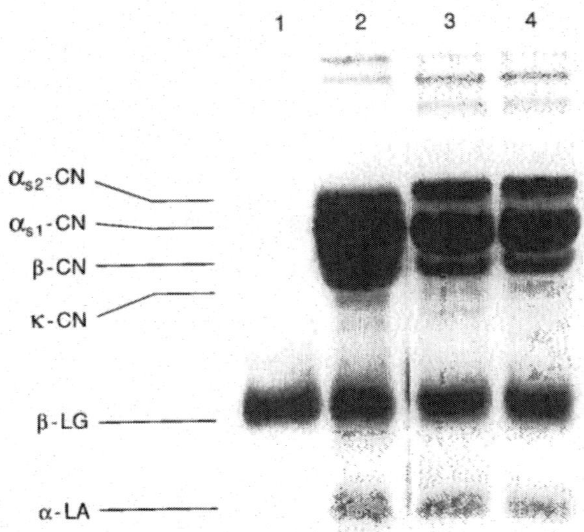

Figure 2. SDS-PAGE photographs of goat milk samples (lanes 3 and 4), cow milk (lane 2), and β-lactoglobulin (lane 1) (Adapted from Guo (2003)).

(1) α_{s1}-Casein is abundant in bovine milk. Until the late 1980s, people held the point that goat milk lacked α_{s1}-casein, a protein homologous to bovine α_{s1}-casein (Jenness, 1980). However, current investigations point that this casein does exist in the milk of certain goat milk with a significantly lower level, and the amount and genetic variants are different between goat breeds (Park et al., 2007). The genetic polymorphisms result in great individual quantitative variations of α_{s1}-casein (Grosclaude, Mahé, Brignon, Di

Stasio, & Jeunet, 1987). More than 18 alleles are identified in goat milk and can be divided into three groups: strong alleles, weak alleles, and null alleles, associated with the level of α_{s1}-casein in milk (Devold et al., 2011; Selvaggi et al., 2014). These differences are depending on DNA and RNA in goat milk (Almaas et al., 2006).

In cow milk, α_{s1}-Casein and β-lactoglobulin are important allergens. Goat milk is regarded as a substitute for cow milk to "cow milk-allergic individuals" according to its nitrogen distribution (Johansson, 2011). Because it contains a lower level of α_{s1}-casein, and certain treatments can partially eliminate the antigenicity of β-lactoglobulin (Tomotake et al., 2006). But some works also noted that it would be risk for people who are allergic to cow milk proteins to ingest goat milk due to cross-allergenicity between goat and cow milk (Pessler, Nejat, & Immunology, 2004; Spuergin et al., 1997). Goat milk as an alternative for people allergic to cow milk is further discussed in chapter 7 "goat milk nutrition and health benefit."

The α_{s1}-Casein can influence digestibility aspects. The α_{s1}-casein is only partly digested by gastric juice enzymes and not totally hydrolyzed until it reaches duodenal enzymes. As a result, the higher amount of α_{s1}-casein in the milk will cost longer time to digest (Almaas et al., 2006). Besides, α_{s1}-casein in goat milk is softer and more easily broken down. Therefore, goat milk is more digestible by the protease in the stomach due to low level and softer α_{s1}-casein (Ceballos et al., 2009). In the stomach, the casein hydrolysis capability of goat milk (about 96%) is higher than that of cow milk (76-90%), but little lower than human milk that can be completely hydrolyzed (Mwenze & Management, 2015). And goat milk takes less time to be digested in a baby's stomach, in 20 minutes, compared to cow milk (2-3 hours) (Attaie & Richter, 2000). The studies show that digestibility is also influenced by heat treatment. Prior heating of cow and goat milk is more resistant to hydrolysis than raw milk (Almaas et al., 2006). Therefore, Goat milk is considered as a better alternative for breast feeding than cow milk concerning better digestibility and nutrient absorption (Mwenze & Management, 2015).

The amount of α_{s1}-casein affects flavors, coagulation and cheese making properties (Slačanac et al., 2010). Goat milk with low level of α_{s1}-CN has lower curd yield, weaker curd firmness, longer rennet coagulation time and more heat liability.

(2) The α_{s2}-casein has been identified from a wide number of species such as caprine and bovine. However, human milk hasn't found α_{s2}-casein (Boisnard & Petrissant, 1985; Selvaggi et al., 2014; Stewart et al., 1987; Martin, Brignon, Furet, & Leroux, 1996). The α_{s2}-casein comprises up to about 29% of the casein fraction in mammalian milk. And α_{s2}-casein in higher concentrations was found in larger casein micelles of caprine (Tziboula & Horne, 1999). The gene encoding of α_{s2}-casein (CSN1S2) has a signal sequence highly conserved which is as disparate group as α_{s1}-caseins (Farrell Jr. et al., 2004). The α_{s2}-caseins are highly phosphorylated peptides and have found at least eight alleles related to three systhesis levels (A, B, C). It contains major component (full length of 207 residues) and several minor components (lacks the internal sequence), displaying varying levels of post-translational phosphorylation as well as minor degrees of intermolecular disulfide bonding (Selvaggi, Laudadio, Dario, & Tufarelli, 2014). Recently, there are two new truncated subvariants were identified in goat α_{s2}-casein (Ferranti, Lilla, Chianese, & Addeo, 2000).

(3) β-casein exists in goat milk as the most abundant protein, which accounts for about 58% of the total casein content. β-casein is sensitive for calcium, even low concentrations of Ca^{2+} can lead to its precipitation. It plays an important role in the surface properties of casein micelles which is necessary for curd formation (Pearse, Linklater, Hall, & Mackinlay, 1986). The studies show that milk without β-casein have lower curd firmness, longer coagulation times and less cheese yield than normal milk (Chianese et al., 1993). The β-casein contains two marsupial proteins and several eutherian and the amino acid and nucleotide sequence are known (Selvaggi et al., 2014). There are more various genetic variants in β-casein than the other caseins and their differentiation is complicated by gel electrophoresis (Whitney et al., 1976).

(4) κ-casein differs from the other casein in that it is soluble in the presence of Ca^{2+} and has a much smaller phosphate component. The κ-casein is very sensitive to enzyme that it has the power to "protect" casein micelle. During the process of rennet-induced coagulation, firstly, κ-casein

(approximately 90%) is rapidly hydrolyzed by enzymatic phase (Walstra & Jenness, 1984). After proteolysis of κ-casein, the casein micelle turns into unstable and susceptible to precipitation by Ca^{2+} and aggregation of the casein micelles occurs (Garnot & Olson, 1982; Walstra & Jenness, 1984). Only κ-casein is identified all sequence of amino acids in the goat milk (Dayhoff & structure, 1979). κ-casein is a longer chain that has 171 amino acids residues than its bovine counterpart. In goat κ-casein, Val and His are inserted at positions 132 and 133, respectively. Rennin (chymosin) hydrolyzes the goat κ-casein between Phe in position 105 and Met in 106, like the bovine homolog, producing the fragments called para-κ-casein (residues 1 to 105) and caseinomacropeptide (residues 106 to 171). There are 19 differences in sequence between cow and goat caseinomacropeptides and only 8 between the para-κ-caseins (Chobert, Mercier, Bahy, & Hazé, 1976; Mercier & Chobert, 1976).

Serum (Whey) Protein

Serum (whey) proteins consist of β-lactoglobulin (β-lg), α-lactalbumin (α-la), serum albumin, immunoglobulins and proteose-peptone fraction. Albumins (β-lactoglobulin and α-lactalbumin), a valuable source of bioactive components with extra physiological activity, make up almost 75% of total whey protein of milk (Selvaggi et al., 2014). Whey protein is typical globular protein resulting from its secondary and tertiary structures, whereas casein lacks tertiary structure. α-helix motifs in whey protein is substantial, which distribute the acidic/basic and hydrophobic/hydrophilic amino acids in a fairly balanced way along their polypeptide chains (Evans, 1982). They are denatured during moderate heating, which may cause their aggregation mainly with the casein micelles (Tsakalidou & Papadimitriou, 2016).

In terms of nutrition, it is mentioned that milk whey protein is superior to casein in many aspects. They is similar to those of human milk in amino acid profile. And whey protein has much better digestion and absorption than casein (Sindayikengera & Xia, 2006). Therefore, whey proteins are recommended for the formulation of infant formulas (Hambraeus, 1982). Whey proteins have many important functional properties such as viscosity, solubility, gelation, water-holding capacity, foaming, emulsification and adhesion, and are widely used as food ingredients (Selvaggi et al., 2014). Different genetic variants of goat whey protein have not been studied in depth yet.

But there are two identified β-lactoglobulin variants and three of α-lactalbumin in goat milk (Moatsou et al., 2005). However, there are not many studies about their physiological importance, but α-lactalbumin required for lactose synthesis is known (Johansson, 2011).

β-Lactoglobulin

β-Lactoglobulin is the primary whey protein in most mammalian milk but is absent in human milk. The β-lactoglobulin is a globular protein member of the lipocalin family. It has the ability to bind small hydrophobic molecules.

Although β-lactoglobulin doesn't yet define the clear physiological function, it is suggested on the transportation of fat-soluble nutritional compounds (Flower, 1996; Pérez & Calvo, 1995). The β-lactoglobulin structure of goat milk is different from that of cow milk and can be distinguished immunologically by microcomplement fixation (Jenness, 1980). However, the sequences of goat β-lactoglobulin and bovine β-lactoglobulin (BLG) have high similarity, with only six positions differences (Loch et al., 2015).

α-Lactalbumin

α-Lactalbumin, one of the major whey proteins in ruminant milks, belongs to the calcium metalloprotein family and owns a single strong Ca^{2+} binding site (Jenness, 1982). α-Lactalbumin exists in all mammalian milk and plays an important role in the synthesis of lactose.

It interacts with the enzyme galactosyl-transferase to synthesize lactose from galactose and glucose in the mammary gland (Ebner & Brodbeck, 1968; Kuhn, 1983). α-lactalbumin moves into mammary secretory vesicles from the inner surface of the mammary Golgi apparatus, then into the alveolar lumen during milk formation (Lönnerdal & Lien, 2003). Goat α-lactalbumin can be distinguished from other α-lactalbumin by using microcomplement fixation technique or by absorption of antibodies by columns of matrix-bound α-lactalbumin, according to immunological cross-reactivity between goat α-lactalbumin and other α-lactalbumin (Prieels et al., 1975). The conformations between goat α-lactalbumin and bovine homolog are similar by analytical method. Tyr, Trp and Lys groups were exposed equally in the conformation of the two proteins (Jenness, 1980).

Amino Acid Composition

The contents of amino acid (mg/100 g milk) in goat milk are shown in Table 3 compared with those in cow milk. It shows that goat milk has greater amount of the amino acids than cow milk, except Tyr and Ser (Ceballos et al., 2009). This is mainly because the higher level of total solids in goat milk, associated with the greater quantity of protein in the dry mater (25.64% verse 24.82%) (Ceballos et al., 2009).

The amino acid composition of each protein fraction in the milk is decided by the species (Marchalonis et al., 1971). Amino acid compositions are found the certain similarities in accordance with each corresponding protein fraction. α_{s2}-casein in goat milk is a polypeptide chain of 217 amino acid residues instead of 207 in cow milk. β-Casein, the most abundant casein fraction in goat milk, includes 213 amino acid residues, whereas bovine β-casein consists of 209 amino acid residues. Goat κ-casein differs from its bovine homolog in that it contains a chain of 171 amino acid residues instead of 169 because Val and His are inserted at positions 132 and 133, respectively.

The hydrolysis site of goat κ-casein by rennet is between Phe and Met at positions 105 and 106 separately which is as same as the bovine counterpart. This process produces para-κ-casein and casein macropeptides and leads to rennet-induced coagulation during cheese-making. Goat β-lg and α-la, both like its bovine counterpart, contain 162 amino acid residues and 123 amino acid residues, respectively (Guo, 2003).

Table 4 shows the amino acid compositions of these proteins. In terms of α-caseins, goat milk presents much higher levels of Asp and Lys than cow milk, but much lower levels of Gly, Ile, Leu and Val. And most the amounts of amino acid residues in goat β-Casein are lower than cow β-Casein, except Gly, Ser, Thr and Ala.

In goat milk, the β-lg contains significantly greater Ala, Met, Pro, Glu and Val than α-la, while the opposite trend is showen for Asp, Cys and Typ (Park & Haenlein, 2008).

Table 3. Amino acid composition (mg/100 g milk) of goat milk and cow milk (n = 30)

	Goat milk	Cow milk
Essential amino acids		
Thr	138.7	115.8
Ileu	160.5	128.0
Leu	341.0	266.2
Lys	342.9	252.6
Met	78.0	71.2
Cys	30.6	23.2
Phe	175.5	133.5
Tyr	162.5	160.0
Val	210.2	147.8
Total	1639.8	1298.4
Nonessential amino acids		
Arg	135.7	114.4
His	122.7	93.1
Ala	118.0	96.1
Asp	250.2	214.2
Glu	694.6	554.3
Gly	55.8	49.2
Pro	310.6	253.4
Ser	152.7	147.9
Total	1840.2	1522.6

Adapted from Ceballos et al. (2009).

Table 4. Amino acid composition of isolated proteins of goat milk and cow milk (g/100 g protein)

	α-Caseina		β-Casein		κ- Caseinb	γ-Caseinc	β-Lactoglobulin		α-Lactalbumin	
Amino acid	Goat	Cow	Goat	Cow	Goat	Cow	Goat	Cow	Goat	Cow
Ala	2.8	3.8	2.4	2	9.4	2.3	9.9	7	4.1	2.1
Arg	3.7	4.3	1.4	3.4	2.9	1.9	1.9	2.8	0.8	1.2
Asp	7.7	0.8	4.2	4.9	9.4	4	8.6	11.4	17.9	18.7
Cys	0.8	0.4	0	0	1.8	0	3.1	3.4	6.5	6.4
Glu	22.9	22.5	20.2	23.2	15.2	22.9	14.8	19.3	10.6	12.9
Gly	0.9	2.3	2.8	1.6	0.6	1.5	3.1	1.4	4.1	3.2
His	2.7	2.9	2.4	3.1	2.3	3.7	1.2	1.6	2.4	2.9
Ile	4.9	6.4	4.2	5.5	6.4	4.4	6.2	6.9	6.5	6.8
Leu	5.3	7.9	9.4	11.6	4.7	12	13.0	15.5	10.6	11.5
Lys	11.1	8.9	5.6	6.5	4.7	6.2	9.9	11.8	10.6	11.5
Met	2.1	2.5	2.8	3.4	0.6	4.1	2.5	3.2	0	0.9
Phe	4.6	4.6	4.2	5.8	2.3	5.8	2.5	3.5	3.3	4.5
Pro	6.9	7.5	15.5	15.1	11.1	17	4.9	5.1	1.6	1.5
Ser	4.8	6.3	7.0	6.8	7.6	5.5	3.7	4	4.9	4.8
Thr	5.6	4.9	5.6	5.1	8.8	4.4	4.9	5	4.9	5.5
Typ	1.5	2.2	0.5	0.8	0.6	1.2	1.2	2.7	3.3	7
Tyr	7.1	8.1	1.9	3.2	5.3	3.7	2.5	3.7	3.3	5.4
Val	4.7	6.3	9.7	10.2	6.4	10.5	6.2	6.1	4.9	4.7

[a] α-Casein represents αs2-Casein for goat milk, and total αs-Casein for cow milk. [b] κ-Casein for goat milk only. [c] γ-Casein for cow milk only.
Adapted from Park & Haenlein (2008).

Non-Protein Nitrogen

Non-protein nitrogen (NPN) are a minor fraction of milk serum, which are various nitrogenous substances in relatively small concentrations. The whole NPN are 5.8% of total nitrogen in bovine milk. They consist of free amino acids, small peptides, creatine, creatinine, urea, ammonia, uric acid, orotic acid, etc (Tsakalidou & Papadimitriou, 2016). The NPN levels in goat and human milk are much higher than those in cow milk. And NPN concentrations vary from different breeds. For example, Alpine goat has lower NPN levels than Nubian, and Hostein cow has lower NPN than Jersey (Park, 1991).

Free amino acids (FAA) constitute 9-10.5% of the NPN in goat milk compared to 16% in sheep milk, 10–20% in cow milk (Renner et al., 1989; Tripaldi et al., 1998). They are more easily absorbed than protein-bound amino acids. The prime free amino acids comprise the nonessential amino acids Gly, Asp, Glu, and Ala, and other free amino acids in very low levels (Renner et al., 1989). Taurine and carnitine, only presenting as free amino acids, are necessary for newborns according to their essential physiological functions. Taurine is the richest FAA in goat milk and the second one in human milk, existing in various mammalian tissues. And taurine is one of the final metabolic products of sulfur-containing amino acids (Tsakalidou & Papadimitriou, 2016). Furthermore, taurine contributes to the digestion and absorption of lipids by forming bile salts (Renner et al., 1989).

Lipids

Lipids are esters of fatty acids and related compounds soluble in non-polar organic solvents and practically insoluble in water. They are classified as three groups: (1) polar lipids (phospholipids and glycolipids); (2) neutral lipids (triacylglycerol, diglyceride and monoglyceride (Chilliard et al., 2003)); and (3) miscellaneous lipids (sterols, carotenoids, and fat-soluble vitamins) (Tsakalidou & Papadimitriou, 2016). In dairy products, lipid composition plays an important part in goat milk nutrition physical and sensory characteristics (Slačanac et al., 2010). In goat milk, the main lipids are free about 97-99% of total milk fat and minor lipids are bound lipids (1-3%) (Table 5).

Free lipids contain 96.8% triacylglycerols (TAG, constituting a very complicated mixture since their fatty acids vary in chain length and solubility), 2.2% diglycerides and 0.9% monoglycerides. Bound lipids consist of 46.8% neutral lipid, 8.5% glycolipid, and 44.7% phospholipid. In this respect, goat milk resembles cow milk (Cerbulis et al., 1982). Lipids also play a role in cheese yield, cheese texture, as well as in the flavor and color of goat dairy products (Delacroix-Buchet & Lamberet, 2000). There are many factors affecting the concentration of milk fat, such as individual traits, breed, nutrition, and stage of lactation (Markiewicz-Kęszycka et al., 2013).

Table 5. Distribution of lipids in free and bound fractions of goat milk

		%
Lipids free	Triglyceride	96.8
	Diglyceride	2.2
	monoglyceride	0.9
Bound lipids	Neutral lipid	46.8
	Glycolipid	8.5
	Phospholipid	44.7

Adapted from Cerbulis et al. (1982).

Fat Globules

The lipids presenting in milk are globule in shape. The concentration of phospholipids in goat milk is 8-10 mg/g fat or 30-40 mg/100 g milk. Phospholipids of goat milk are mainly distributed in fat globule membrane, as well as around 40% of total phospholipids in the skim milk (Guo, 2003). Goat milk is viewed as "natural homogenized" milk. For size distribution of fat globules, the proportion of smaller particles in goat milk are much larger than in cow milk. And the average diameter of goat lipid globules is the smallest among ruminants except camel (Barłowska et al., 2011; Johansson, 2011). The average diameter of goat milk lipid globules is around 2.5 μm as compared to 3 μm for cow milk fat and around 4 μm for human milk (Tsakalidou & Papadimitriou, 2016). And after *pasteurization,* the size of milk lipid globule doesn't change significantly (Ma et al., 2019).

Total fat content and fat globules size and distribution affect the physicochemical and sensory characteristics of dairy products and are of importance for industrial significance; it also affects *cheese ripening,* as well as the shelf life of dairy products (Attaie & Richter, 2000; Michalski et al., 2003). Additionally, goat milk has a poor creaming ability at cooler temperatures due to the absence of agglutinin, a substance that gives rise to the clustering of the fat globules (Jennes & Parkash, 1971).

In goat milk, a greater proportion of small diameter of fat globules leads to a more homogeneous mixture and better dispersion of fat. Small milk fat globules have a relatively larger total surface area of the globules compared to large globules which may make goat milk more digestible (Argov-Argaman, 2019; Park & Haenlein, 2008). Greater content of membrane makes lipids more effective to contact with enzymes in the gastrointestinal tract with pancreatic lipase activity (Attaie & Richter, 2000; López-Aliaga et al., 2010) (Table 6).

Table 6. Mean diameter (dvs), range, and specific surface area (SSA) of particles in milk

	Whole milk			Skim milk	
	dvs (μm)	Range (μm)	SSA (cm2/ml)	dvs (μm)	SSA (cm2/ml)
Bovine	3.51	0.92-15.75	17,117	0.213	282,100
Caprine	2.76	0.73-8.58	21,778	0.225	266,411

Adapted from Attaie and Richter (2000).

Fatty Acid

The fatty acids (FA) could be divided into various groups according to their chemical structure: short chain, medium chain and long chain, saturated, monounsaturated and polyunsaturated, cis and trans, branched, conjugated. Fatty acids are regarded as potential positive or negative predisposing factors associated with the health of human consumers (Williams, 2000). The fatty acids used to synthesize the milk triacylglycerols arise from two sources: from plasma lipids and by de novo synthesis generally involves short-chain fatty acids ($C_{4:0}$-$C_{12:0}$) and some $C_{16:0}$ (Tsakalidou & Papadimitriou, 2016). Most carbon atom numbers of fatty acids are even, like from acetic ($C_{2:0}$) to arachidic acid ($C_{20:0}$) (Park et al., 2007). Different fatty acids have different esterification sites, for example, most short-chain fatty acids are esterified at the 3-position of glycerol and longer chains are at 1 or 2 position. And ester linkages of short-chain fatty acids are more easily attacked by lipases compared to those of longer fatty acids (Guo, 2003).

Table 7. Fatty acid profiles of goat, cow and human milk fat (%)

Fatty acids	Goat	Cow	Human
C4	0.7	1.4	0.1
C6	2.4	2.2	0.2
C8	3.2	1.8	0.3
C10	8.7	3.6	2.1
C12	4.7	4.0	7.2
C14	10.7	13.0	10.9
C16	28.5	30.2	29.6
C18	13.0	13.7	7.3
C18:1	25.2	27.1	35.3
C18:2	2.9	3.0	6.7

Adapted from: Guo (2003).

Saturated fatty acids (SFA) are the prime components of milk fat acids, in range of 67-75% (Rodríguez-Alcalá et al., 2009; Žan et al., 2006). Goat milk fat has a relatively higher amounts of short and medium chain fatty acids in total lipids than cow and human milk fat (Table 7), especially caproic ($C_{6:0}$), caprylic ($C_{8:0}$), capric ($C_{10:0}$), lauric ($C_{12:0}$) acids (Park & Haenlein, 2008). Among these fatty acids, caproic, caprylic and capric acids, as predominance in goat milk, have actually been named after caprine, which are synthesized de novo in the mammary gland (Chilliard, et al., 2006; Haenlein, 2004). Goat milk has higher levels of caproic ($C_{6:0}$), caprylic ($C_{8:0}$), capric($C_{10:0}$) acids than cow milk, whose free state are highly correlated to "goaty" flavor (Park & Haenlein, 2008).

Goat milk has a low level of short and medium-chain fatty acids (C_4-C_{14}) and a high level of $C_{18:0}$ and $C_{18:1}$ which is drawing on their body fat at the beginning of the lactation. The long-chain fatty acid profile of goat milk resembles that of cow milk (Guo, 2003). Interestingly, the ratio of lauric: capric acid (C_{12}: C_{10}) is special in the goat milk cheese, since it is significantly lower than that in cow milk cheese (0.46 versus 1.16) (Iverson & Sheppard, 1989). And the ratio increases proportionally as increasing substitution of cow milk in place of goat milk in the manufacture of cheeses. It has been used in the dairy industry to detect the extent of adulteration of goat milk with cow milk or cheese. The $C_{16:0}$ and $C_{18:1}$ acids represent in goat and cow milk fats at markedly high concentrations and the ratio between them is no species-specific (Park & Haenlein, 2008).

Branch-chain fatty acids (BCFA) and odd-number chain saturated fatty acids may contribute characteristic flavors to many dairy foods. Their amounts in cheese depend on the composition of the milk fat. There are 36 kinds of BCFA (mostly monomethylates) identified and quantified in goat milk (Alonso et al., 1999). Table 8 showed 30 minor BCFA detected by gas chromatography/ mass spectrometry (GC/MS), ranging from 0.004 to 0.765 mg/g of total FA (Alonso et al., 1999). These identified BCFA consist of 25 monomethyl BCFA, 2 dimethyl BCFA, and 4 ethyl BCFA. 4-Ethyloctanoate as one of Ethyl BCFA especially existed in goat milk and was absent in cow milk. 4-Ethyloctanoate and 4-methyloctanoate represented goaty and muttony flavors of dairy products. Other BCFA like 3-methylbutanoate, 4-methylpentanoate, and 8-methylnonanoate were detected in both caprine and bovine milk and also influence the characteristic flavor of dairy foods (Ha & Lindsay, 1991). The main methyl substitution of C_4 and C_6 in goat milk is monomethyl-branched components, other than iso- or anteiso- fatty acid, which is no found in bovine milk.

Table 8. Composition and tentative identification of minor branched-chain and odd-numbered chain saturated fatty acids (mg/g of total methyl esters)

Carbon number	Identification	$\bar{X}$
5	2-Methylbutanoate	0.005
5	3-Methylbutanoate	0.023
6	2-Ethylbutanoate	0.004
5	Pentanoate(C5:0)	0.134
6	3-Methylpentanoate	0.025
6	4-Methylpentanoate	0.017
8	2-Ethylhexanoate	0.007
7	4-Methylhexanoate	0.05
7	Heptanoate(C7:0)	0.363
8	Methylheptanoate	0.026
9	2,4-Dimethylheptanoate	0.051
8	Methyloctanoate	0.028
9	4-Methyloctanoate	0.391
9	6-Methyloctanoate	0.041
9	Nonanoate(C9:0)	0.915
10	4-Ethyloctanoate	0.227
10	4-Methylnonanoate	0.016
10	8-Methylnonanoate	0.036
11	Methyldecanoate	0.66
11	Methyldecanoate	0.016
11	Methyldecanoate	0.027
12	2-Ethyldecanoate	0.322
11	Undecanoate(C11:0)	0.233
12	Dimethyldecanoate	0.31
13	4-Methyldodecanoate	0.135
14	12-Methyltridecanoate(iso)	0.765
14	11-Methyltridecanoate(anteiso)	0.37
15	Methyltetradecanoate	0.348

Table 8. (Continued)

Carbon number	Identification	$\bar{X}$
15	Methyltetradecanoate	0.23
16	13-Methylpentadecanoate(anteiso)	0.131
17	Methylhexadecanoate	0.25
17	Methylhexadecanoate	0.243
17	Methylhexadecanoate	0.197
18	16-Methylheptadecanoate	0.578
19	Nonadecanoate(C19:0)	0.161
21	Methyleicosanoate	0.322

Adapted from Alonso et al. (1999).

Unsaturated Fatty Acids

Monounsaturated fatty acids (MUFA) comprise around 20-32% of the total fatty acids in the goat milk fat (Žan et al., 2006; Talpur et al., 2009). For major mammals, the typical character is that oleic acid ($C_{18:1}$) is the highest component among the MUFA family and its level increased with the progress of lactation (G. F. Haenlein & Wendorff, 2006; Strzałkowska et al., 2009; Tsiplakou & Zervas, 2008).

Polyunsaturated fatty acids (PUFA) refer to the carbon backbone of fatty acids containing more than one double bond, ranging from 3% to 5% of a total FA pool. PUFA are the most valuable group among FA because of their favorable effect on the health of consumers (Rodríguez-Alcalá et al., 2009). Goat and cow milk fat can provide sufficient amounts of essential fatty acids for human infants. Goat milk has higher proportions of ω-3 and ω-6 polyunsaturated fatty acids (PUFA), as well as conjugated linoleic acid (CLA) (Ceballos et al., 2009). During the progress of lactation, the content of PUFA increased and reached about 4.5% (Strzałkowska et al., 2009). Nowadays, studies show that the levels of PUFA in milk can be enhanced by increasing PUFA oil intake in animals' dietary or decreasing their hydrogenation in the rumen. It has been proved these methods are successful for goat milk, such as high forage/concentrate ratio, and trapping of FA in vegetable cells (Chilliard & Ferlay, 2004).

Milk unsaturated fatty acids may comprise one or several trans double bonds. Trans fatty acid occurs in milk partly due to rumen microbial activity. PUFA from the diet, like linoleic ($C_{18:2}$) and linolenic ($C_{18:3}$) acids, metabolize by rumen flora to stearic acid ($C_{18:0}$) via forming trans-$C_{18:1}$ acid. Then conjugated linoleic acid and trans-$C_{18:1}$ acid formation, which lead to linoleic acid biohydrogenation producing stearic acid (Chilliard et al., 2000; Griinari and Bauman, 1999). The average content of trans-$C_{18:1}$ isomer in goat milk was about 1.5-2.1% (relative to total fatty acids), while those in cow milk was about 2.5-3.8% (Alonso, Fontecha, Lozada, Fraga, & Juárez, 1999; LeDoux, Rouzeau, Bas, & Sauvant, 2002). The distribution of each trans-C18:1 isomer observed in goat milk fat is similar to those in cow milk fat (Precht, 1995). Trans-vaccenic acid (trans-11-$C_{18:1}$) was the major component in goat and cow milk and its double bond was located on carbons 11 (35-40%) (Alonso et al., 1999; Bickerstaffe et al., 1972; LeDoux et al., 2002), while human milk fat trans $C_{18:1}$ mainly located on carbons 6 to 14. The studies showed that trans-$C_{18:1}$ fatty acid increased with decreasing the level of forage in the diet (LeDoux et al., 2002).

Conjugated linoleic acid (CLA) contains all positional and geometric isomers of linoleic acid with a conjugated double bond (Park et al., 2007). CLA primarily comes from the desaturation of vaccenic acid by the delta-9 desaturase enzyme (Chilliard et al., 2003). In general, goat milk has greater proportion of CLA than cow milk (Ceballos et al., 2009). Under similar diets, the *trans*-vaccenic and rumenic acid contents in goat milk fat were lower than those in cow milk fat. Monogastric farm animals such as mare or sow which do not feed on margarine or ruminant milk fat almost don't contain tr*ans*-vaccenic acid and rumenic acid (Jahreis et al., 1999).

It is evidenced that CLA has diverse beneficial properties via many animal studies and gained great attention in human nutrition in recent years such as anti-obesity and anticarcinogenic effects (Williams, 2000). It is a viable way for domestic animals to change the fatty acid composition and increase CLA content in their milk via dietary manipulation, which may be in favor of human health. CLA content in dairy products varies in many factors, and variations of feeding conditions in different seasons are one of the factors changing milk CLA concentration in different ruminant species

(Chilliard & Ferlay, 2004). The studies showed that dairy goats and ewes feeding oilseed or oil supplements rich in PUFA have been effective in enriching milk (Chilliard et al., 2005; Luna et al., 2005). For example, canola oil is a great oil supplement for the goat diet which had been proved effectively increasing the content of CLA and essential fatty acids in goat milk (Guo, 2003), whereas feeding goat with whole untreated *oil seeds* didn't change significantly for goat milk.

Sterol Fraction

Milk lipids also contain sterols in relatively small concentrations. Cholesterol is the main sterol in milk, most in the free form, with only a small fraction in the form of esters. The content of cholesterol in fresh goat milk is in the range of 0.10-0.2 mg/ml, while that in the goat colostrums is much higher (0.40 mg/ml) (Guo, 2003). It is believed that cholesterol in plasma is concerned with cardiovascular disease. In milk and dairy products, analyzing sterol fractions is used to detect vegetable fats of adulteration. The concentrations of cholesterol in goat and sheep milk were changeable according to different reports, because of different breeds, as well as different analytical techniques (Park et al., 2007).

Carbohydrates

The main carbohydrate in goat milk is lactose. The amount of lactose in goat milk (4.2-4.8%) is a little lower compared to that in cow milk (Guo, 2003). Lactose consists of a glucose and a galactose molecule. This disaccharide is synthesized with the required active participation of the milk protein α-lactalbumin in the mammary (Kunz, et al., 2000). Lactose is of nutritional interest, as it is beneficial to intestinal absorption and utilization of certain minerals and vitamins, such as calcium (Ca), magnesium (Mg) and phosphorus (P) and vitamin D (Campbell & Marshall, 1975). Lactose plays an important part in maintaining osmotic equilibrium between the bloodstream and the alveolar cells of the mammary gland during milk synthesis, and secretion into the alveolar lumen and the duct system of the udder (Larson & Smith, 1974). Goat milk may be easier to be digested. This is helpful for people suffering from unhealthy intestinal functions such as intestinal bowel syndrome (Mwenze & Management, 2015).

Other minor carbohydrates found in goat milk are oligosaccharides, glycoproteins, glycopeptides and nucleotide sugars (Larson & Smith, 1974). The oligosaccharides are the linear or branched chains made up of 3 to 10 monosaccharides in milk (Tsakalidou & Papadimitriou, 2016). The content of lactose-derived oligosaccharides in goat milk is in that range of 0.25-0.3g/L, higher than that in bovine and ovine milk, but significantly lower than that in human milk (5-16g/L) (Lad et al., 2017; Tsakalidou & Papadimitriou, 2016). The superiority of goat milk oligosaccharide is the most similar structure to human milk oligosaccharides that are conducive to infant health according to their anti-infective and prebiotic properties (Lara-Villoslada et al., 2004). The studies showed that goat milk oligosaccharides have anti-inflammatory effects in induced colitis by animal models (Daddaoua et al., 2006). These results could be valuable in the management of inflammatory bowel disease (Robinson, 2001).

Nucleotide sugars in milk are the precursors of glycoproteins, oligosaccharides, and glycolipids in the biosynthesis of milk and the glycosyl donors for glycosyltransferase in milk and mammary gland. Normal goat milk contains the highest nucleotide content, 154 μmol/100 ml, among sheep (93 μmol/100 ml), cow (68 μmol/100 ml) and human milk (5 μmol/100 ml). Nucleotide concentrations in colostrums in these milks are 271, 499, 58 and 6 μmol/100 ml, respectively (Park et al., 2007).

Vitamins

Vitamins are divided into water- or fat-soluble vitamin according to their solubility. Vitamin B-complex and vitamin C (ascorbic acid) belong to the water-soluble group, while fat-soluble vitamin comprises vitamin A, D, E and K (Tsakalidou & Papadimitriou, 2016). Vitamins composition for different species are shown in Table 9. Vitamin A is the most abundant vitamin in goat milk and its content is similar to that in human milk, higher than cow milk. This is because goats transform total β-carotene from foods into vitamin A in the milk (Conesa et al., 2008). Goat milk always looks whiter than cow milk for the same reason. Vitamin A is of importance for both innate and adaptive immune responses, including cell-mediated immunity and antibody responses. Goat milk supplies adequate amounts of vitamin A and niacin, and excesses of pantothenate thiamin and riboflavin for human (Lad et al., 2017; Tsakalidou & Papadimitriou, 2016). However, if a human infant only fed with goat milk instead of human milk, the infant may be oversupplied with protein, Ca, P, vitamin A, pantothenate, riboflavin, niacin and thiamin according to the FAO-WHO requirements (Jenness, 1980).

Table 9. Vitamin contents of goat, cow and human milk (per 100g)

	Goat	Cow	Human
Vitamin A (IU)	185	126	190
Vitamin D (IU)	2.3	2	1.4
Thiamine (mg)	0.068	0.045	0.017
Riboflavin (mg)	0.21	0.16	0.02
Niacin (mg)	0.27	0.08	0.17
Pentathenic acid (mg)	0.31	0.32	0.2
Vitamin B6 (mg)	0.046	0.042	0.011
Folic acid (μg)	1	5	5.5
Biotin (μg)	1.5	2	0.4
Vitamin B12 (μg)	0.065	0.357	0.03
Vitamin C (μg)	1.29	0.94	5

Adapted from Guo (2003) and Park et al. (2007).

Vitamin B group in milk comes from rumen synthesis and somewhat independent of diet (Haenlein & Caccese, 1984; Mann, 1988). In goat milk, the contents of folic acid and vitamin B12 are five times lower than those in cow milk and the significant deficiencies in folic acid and B12 can cause "goat milk anemia" (Park et al., 2007). Low level of folic acid, as the hemoglobin synthesis necessary, is the main factor responsible for "goat milk anemia" in goat milk, although vitamin B12 deficiency can cause megaloblastic anemia in infants. Baby diet must supplement some vitamins into goat and cow milk due to both milk deficient in pyridoxine (B6), Vitamins C and D (McClenathan & Walker, 1982). High-temperature short-time pasteurization (HTST) is recommended as the best sterilization method of goat milk to preserve vitamins as well as to extend shelf-life of the milk, in spite of the occurrence of some thiamine, riboflavin and vitamin C losses (Lavigne et al., 1989).

Minerals

Goat milk has abundant minerals and is considered as an exceptionally important food for human nutrition. Mineral composition of the milk may be used to distinguish between the goat and cow milk because the milk of each species has a particular individual pattern of minerals, which may be related to the nutritional importance of the element (Rincón et al., 1994). The contents of macrominerals may fluctuate little, but they still are affected by the breed, individual animal, diet, status of udder health and stage of lactation (Park & Chukwu, 1988). An inverse correlation between lactose concentration and the molar sum of Na and K concentrations exist in goat milk. Chloride has a positive correlation with K and a negative relationship with lactose, while Na is not significantly associated with K, Cl, and lactose. Substantial changes in the contents of the major minerals in goat milk were found during the first 7 weeks of lactation (Maraval & Vignon, 1982). With the process of lactation, the macro elements decreased: Mg from 0.21 - 0.27 to 0.10 - 0.13, Na from 0.43 - 0.48 to 0.30 - 0.37 g/l, P from 1.43 - 1.57 to 0.90 - 0.93, and Ca from 1.80 - 2.00 to 1.23 - 1.41, separately.

As shown in Table 10, the contents of mineral in goat milk are compared to those in cow and human milk. Goat milk contains higher amounts of Ca, P, K, Mg and Cl (mg/100 g of milk) than cow and human milk. The content of Na and S in goat milk is higher than human milk but lower than cow milk. In terms of human nutrition, goat milk is more valuable than cow milk according to higher level of most of the minerals. As a result, human milk is unique for infant and is not substituted by either goat or cow milk (Lad et al., 2017).

It is widely recognized that Calcium plays an important part in maintaining the density and strength of bones. Due to high Ca content, goat milk is often recommended as a great source of Ca when it comes to minerals. From the viewpoint of total quantity and correlation with P, it is difficult to obtain an adequate intake of Ca, if people don't consume appreciable milk or other dairy products (Ceballos et al., 2009). For nutrition requirement, it is important to know that goat milk provides about 33% Ca and 27% P of dairy value (DV) in one cup, whereas cow milk supplies around 30% Ca and 23% P of dairy value (DV) in the same volume (Mwenze & Management, 2015).

The contents of Ca and P in goat milk is the highest among three milks about 134 mg and 121 mg/100 g, followed by cow milk (122 mg and 119 mg/100g), and human milk (33 mg, 43 mg/100g) (Table 10). In milk, Ca and P participate in the formation of calcium phosphate which is a main component of the mineral complex-hydroxyapatite that gives structure and strength to bones. A study showed that the goat cheese in the diet every day was the best way for utilization of Ca than supplemental Ca (1000 mg) + vitamin D3 (200U), supplemental Ca (1000 mg/day) alone or placebo. Especially, those teenager girls in this experiment had higher whole-body bone mineral density and cortical thickness of the tibia by eating cheese every day. Therefore, goat cheese may be viewed as a valuable supplement for cortical bone mass accrual (Mwenze & Management, 2015). Furthermore, compared with cow milk, goat milk better enhanced Ca content of rats' bones (Aliaga et al., 2000).

Sodium is particularly important for organisms enabling healthy cell performance and proper neurological function and correlated tightly with amino acids. There is only one channel for the absorption of amino acids through the small intestine depending on the sodium-dependent transport system. If the absence of Na, certain amino acids cannot be absorbed. Goat milk can give a hand on amino acid absorption as a Na source (Guo, 2003).

Potassium, as one of electrolytes, is also a crucial mineral playing a multitude of systemic roles. Potassium deficiency is a common phenomenon, which can cause hypertension, fatigue, muscle weakness and depression due to acid-base imbalance in the body. Potassium can be acquired from fruits, vegetables, whole grains and so on. But most Americans still are lack of K because of inadequately intaking these foods. The highest concentration of mineral is K in goat and cow milk. Furthermore, goat milk is a kind of alkaline foods due to the mineral levels and helps keep the intestinal tract from becoming overly acidic (Guo, 2003).

Table 10. Mineral content of ruminant and human milk (mg/100g)

Minerals	Goat	Cow	Human
Calcium	134	122	33
Phosphorus	121	119	43
Magnesium	16	12	3
Potassium	181	152	55
Sodium	41	58	15
Chlorine	150	100	60
Sulfur	28	32	14
Iron	0.07	0.08	0.2
Copper	0.05	0.06	0.06
Manganese	0.032	0.02	0.07
Zinc	0.56	0.53	0.38

Adapted from Park et al. (2007).

Trace elements of goat milk refer to iron (Fe), copper (Cu), manganese (Mn), and zinc (Zn). The contents of trace elements of goat milk vary according to diet, breed, individual animals, and stage of lactation. The colostrum of goat milk has higher concentrations of trace elements than normal milk (Guo, 2003).

Goat milk has comparable levels of trace elements with cow milk, but different from human milk. The content of Fe in goat milk is about 0.07 mg/100 ml, lower than human milk, however, human milk is also deficient in Fe element. The content of Zn is the highest among the trace elements of goat milk, at 0.56 mg/100 ml, higher than human milk.

PHYSICAL PROPERTIES

The physical characteristics of lipids in goat and sheep milk are not much different compared with cow milk. The differences are shown in Table 11. Goat milk contains higher viscosity, specific gravity, and titratable acidity than cow milk, whereas the refractive index and freezing point of goat milk is lower compared to cow milk (Parkash & Jenness, 1968). From the average specific gravity of goat milk was found increased in each lactation (Bhosale et al., 2009). The relationship between percentage of TS and specific gravity (SG) and fat (F) content for goat milk has been reported the first time by Guo's lab (Guo et al., 2001) as follows:

$$TS = 0.13G + 1.41F + 4.28\ (r^2 = 0.94, P < 0.01)$$

Viscosity presents one of the physical properties of fluid, reflecting the relationship between kinetic motion and free surface. Viscosity and specific gravity in goat milk are slightly higher compared to cow milk, much higher than water because of the internal friction like lipids of milk, while water contains relatively higher surface tension than goat milk (Spreer, 1998). It was observed that the viscosity of goat milk significant affected by lactation (Bhosale et al., 2009). Viscosity is changeable according to state and concentrations of fat and protein, temperature, pH and age of the milk (Jenness & Patton, 1974).

Goat milk has the same acid values and an unsaponifiable matter of milk fat with cow milk. Saponification value and refractive index in milk are correlated with longer carbon chains and saturation of the fatty acids, which is the reason for higher those indexes in cow milk rather than goat milk (Park & Haenlein, 2008). The acidity in goat milk is in the form of lactic acid, around 0.18%. The osmotic pressure of goat milk is around 295 mosmol kg-1 H_2O (Guo, 2003). The freezing point of goat milk is a little lower than cow milk indicating a higher level of solute component, such as salts. Goat milk contains higher average electrical conductivity at 0.0091 $\Omega\ cm^{-1}$ than that of cow milk (0.0048 $\Omega\ cm^{-1}$).

Table 11. Comparison of physico-chemical characteristics of lipids and micelle structures between goat, sheep and cow milk

Properties	Goat milk	Cow milk
Specific gravity (density)	1.034	1.031
viscosity, Cp	2.12	2
Surface tension (Dynes/cm)	52	47.2
Conductivity (Ω cm-1)	0.0091	0.0048
Refractive index	1.45	1.451
Freezing point (- oC)	0.556	0.55
Acidity (lactic acid %)	0.18	0.165
pH	6.65	6.68
Unsaponifiable matter of milk fat (%)	0.41	0.41
Acid value	0.47	0.48
Iodine value	19.5	27.09
Saponification value	228.6	232.3
Reichert Meissl value	19.5	29
Polenske value	1.8	1.35
Fat globule diameter (μm)	3.49	4.55
Non-centrifugal casein (% of total casein)	8.7	5.7
Average diameter (nm)	260	180
Hydration of micelle (g/g MS)	1.77	1.9
Mineralization of micelle (g/ca/100 casein)	3.6	2.9

Adapted from Park et al. (2007).

The difference of physicochemical properties between the goat and cow milk is mainly reflected in some different technological behavior of the two milks (Remeuf, 1992). The quality of dairy products including cheese are evaluated according to various rheological/textural properties and casein gels take important roles in gel, stretch and fracture properties. Compared with cow milk, goat milk has a poor cheese-making ability which is demonstrated in more cheese particles lost in whey, lower cheese yield, and likely less textural integrity. Casein characters of goat milk are the major factor of poor coagulum strength of cheese, such as α_{s1}-casein in a low concentration, as well as size, composition and hydration properties of casein micelles (Remeuf, 1992). Studies showed that Goat rennet curd had a shorter coagulation time and more fragile and softer texture than cow counterparts (Abou-Dawood & El-Sawaf, 1977; Remeuf et al., 1989; Puri and Parkash, 1962, Remeuf et al., 1989).

REFERENCES

Abou-Dawood, A. E., & El-Sawaf, S. (1977). Curd tension of milk from some mammals in comparison to some infant milk powders. *Egyptian Journal of Dairy Science,* 5(2), 129-133.

Ahmad, S., Gaucher, I., Rousseau, F., Beaucher, E., Piot, M., Grongnet, J. F., & Gaucheron, F. (2008). Effects of acidification on physico-chemical characteristics of buffalo milk: A comparison with cow's milk. *Food Chemistry*, 106(1), 11-17.

Aliaga, I. L., Alférez, M. J. M., Barrionuevo, M., Lisbona, F., & Campos, M. S. (2000). Influence of goat and cow milk on the digestive and metabolic utilization of calcium and iron. *Journal of Physiology and Biochemistry,* 56(3), 201-208.

Almaas, H., Cases, A. L., Devold, T. G., Holm, H., Langsrud, T., Aabakken, L., & Vegarud, G. E. (2006). *In vitro* digestion of bovine and caprine milk by human gastric and duodenal enzymes. *International Dairy Journal*, 16(9), 961-968.

Alonso, L., Fontecha, J., Lozada, L., Fraga, M. J., & Juárez, M. (1999). Fatty acid composition of caprine milk: major, branched-chain, and trans fatty acids. *Journal of Dairy Science*, 82(5), 878-884.

Argov-Argaman, N. (2019). Symposium review: Milk fat globule size: Practical implications and metabolic regulation. *Journal of Dairy Science,* 102(3), 2783-2795.

Attaie, R., & Richter, R. L. (2000). Size distribution of fat globules in goat milk. *Journal of Dairy Science*, 83(5), 940-944.

Ballester, M., Sanchez, A., & Folch, J. M. (2005). Polymorphisms in the goat β-lactoglobulin gene. *Journal of Dairy Research*, 72(3), 379-384.

Barłowska, J., Szwajkowska, M., Litwińczuk, Z., & Król, J. (2011). Nutritional value and technological suitability of milk from various animal species used for dairy production. *Comprehensive Reviews in Food Science and Food Safety*, 10(6), 291-302.

Bhosale, S. S., Kahate, P. A., Kamble, K., Thakare, V. M., & Gubbawar, S. G. (2009). Effect of lactation on physico-chemical properties of local goat milk. *Veterinary World*, 2(1), 17.

Bickerstaffe, R., Noakes, D. E., & Annison, E. F. (1972). Quantitative aspects of fatty acid biohydrogenation, absorption and transfer into milk fat in the lactating goat, with special reference to the cis-and trans-isomers of octadecenoate and linoleate. *Biochemical Journal*, 130(2), 607-617.

Boisnard, M., & Petrissant, G. (1985). Complete sequence of ovine αs2-casein messenger RNA. *Biochimie,* 67(9), 1043-1051.

Calderon, I., De Peters, E. J., Smith, N. E., & Franke, A. A. (1984). Composition of goat's milk: Changes within milking and effects of a high concentrate diet. *Journal of Dairy Science*, 67(9), 1905-1911.

Campbell, J. R., & Marshall, R. T. (1975). *The science of providing milk for man.* McGraw Hill Book Company.

Ceballos, L. S., Morales, E. R., de la Torre Adarve, G., Castro, J. D., Martínez, L. P., & Sampelayo, M. R. S. (2009). Composition of goat and cow milk produced under similar conditions and analyzed by identical methodology. *Journal of food Composition and Analysis*, 22(4), 322-329.

Ceballos, L. S., Sampelayo, M. S., Extremera, F. G., & Osorio, M. R. (2009). Evaluation of the allergenicity of goat milk, cow milk, and their lactosera in a guinea pig model. *Journal of Dairy Science*, 92(3), 837-846.

Cerbulis, J., Parks, O. W., & Farrell Jr, H. M. (1982). Composition and distribution of lipids of goats' milk. *Journal of Dairy Science*, 65(12), 2301-2307.

Chianese, L., Garro, G., Nicolai, M. A., Mauriello, R., Ferranti, P., Pizzano, R. & Rando, A. (1993). The nature of β-case in heterogeneity in caprine milk. *Le Lait*, 73(5-6), 533-547.

Chianese, L., Portolano, B., Troncone, E., Pizzolongo, F., Ferranti, P., Addeo, F. & Calagna, G. (2000, May). *The quality of Girgentana goat milk*. In Proc. 7th Int. Conf. on Goats, France, 946-949.

Chilliard, Y., Chabosseau, J. M., Rouel, J., Capitan, P., Gominard, C., Gaborit, P. & Ferlay, A. (2002). Interactions between forage nature and sunflower or linseed oil supplementation on goat milk fatty acids of interest for human nutrition. *Proc. 19th Gen. Mtg. of the Eur.* Grassl. Fed. La Rochelle, France, 27-30.

Chilliard, Y., Ferlay, A., Mansbridge, R. M., & Doreau, M. (2000). Ruminant milk fat plasticity: nutritional control of saturated, polyunsaturated, trans and conjugated fatty acids. In *Annales de zootechnie* 49(3), 181-205

Chilliard, Y., Ferlay, A., Rouel, J., & Lamberet, G. (2003). A review of nutritional and physiological factors affecting goat milk lipid synthesis and lipolysis. *Journal of Dairy Science*, 86(5), 1751-1770.

Chilliard, Y., & Ferlay, A. (2004). Dietary lipids and forages interactions on cow and goat milk fatty acid composition and sensory properties. *Reproduction Nutrition Development*, 44(5), 467-492.

Chilliard, Y., Rouel, J., Ferlay, A., Bernard, L., Gaborit, P., Raynal-Ljutovac, K., & Lauret, A. (2005). Effects of type of forage and lipid supplementation on goat milk fatty acids and sensory properties of cheeses. *International Dairy Federation Special Issue*, (5), 297-304.

Chilliard, Y., Rouel, J., & Leroux, C. (2006). Goat's alpha-s1 casein genotype influences its milk fatty acid composition and delta-9 desaturation ratios. *Animal Feed Science and Technology*, 131(3-4), 474-487.

Chobert, J. M., Mercier, J. C., Bahy, C., & Hazé, G. (1976). Structure primaire du caseinomacropeptide des caseines κ porcine et humaine. *Febs Letters*, 72(1), 173-178.

Conesa, C., Sánchez, L., Rota, C., Pérez, M. D., Calvo, M., Farnaud, S., & Evans, R. W. (2008). Isolation of lactoferrin from milk of different species: calorimetric and antimicrobial studies. *Comparative Biochemistry and Physiology Part B: Biochemistry and Molecular Biology*, 150(1), 131-139.

Daddaoua, A., Puerta, V., Requena, P., Martinez-Ferez, A., Guadix, E., Sanchez de Medina, F., & Martinez-Augustin, O. (2006). Goat milk oligosaccharides are anti-inflammatory in rats with hapten-induced colitis. *The Journal of Nutrition*, 136(3), 672-676.

Dayhoff, M. O. (1979). Kappa casein-goat. *Atlas of Protein Sequence and Structure*, 5(3), 301.

Delacroix-Buchet, A., & Lamberet, G. (2000). Sensorial properties and typicity of goat dairy products. *International Conference on Goats*, 7, 559-563.

Destaillats, F., Wolff, R. L., Precht, D., & Molkentin, J. (2000). Study of individual trans-and cis-16:1 isomers in cow, goat, and ewe cheese fats by gas-liquid chromatography with emphasis on the trans-Δ3 isomer. *Lipids*, 35(9), 1027-1032.

Devold, T. G., Nordbø, R., Langsrud, T., Svenning, C., Brovold, M. J., Sørensen, E. S., & Vegarud, G. E. (2011). Extreme frequencies of the α s1-casein "null" variant in milk from Norwegian dairy goats-implications for milk composition, micellar size and renneting properties. *Dairy Science & Technology*, 91(1), 39-51.

Ebner, K. E., & Brodbeck, U. R. S. (1968). Biological role of α-lactalbumin: A review. *Journal of Dairy Science*, 51(3), 317-322.

Evans, E. W. (1982). Uses of milk proteins in formulated foods. In: Hudson BJF, (Ed.), Developments in food proteins. *Appl Science*, London, UK.

Farrell Jr, H. M., Jimenez-Flores, R., Bleck, G. T., Brown, E. M., Butler, J. E., Creamer, L. K., & Swaisgood, H. E. (2004). Nomenclature of the proteins of cows' milk-Sixth revision. *Journal of Dairy Science*, 87(6), 1641-1674.

Ferranti, P., Lilla, S., Chianese, L., & Addeo, F. (2000). Qualitative and quantitative variation of deleted protein forms in goat αs1 and αs2 casein phenotypes. In *Proceedings 7th International Conference on Goats*, Tour, France, 15-18.

Flower, D. R. (1996). The lipocalin protein family: structure and function. *Biochemical Journal*, 318(1), 1-14.

Folin, O., Denis, W., & Minot, A. S. (1919). Lactose, fat, and protein in milk of various animals. *Journal of Biological Chemistry,* 37(3), 349-352.

Fox, P. F., McSweeney, P. L., & Paul, L. H. (1998). *Dairy Chemistry and Biochemistry*. Blackie Academic and Professional, London, UK.

Garnot, P., & Olson, N. F. (1982). Use of oscillatory deformation technique to determine clotting times and rigidities of milk clotted with different concentrations of rennet. *Journal of Food Science*, 47(6), 1912-1915.

Graziano, M., D'Andrea, M., Angiolillo, A., Lagonigro, R., & Pilla, F. (2003). A new polymorphism in goat beta-lactoglobulin promoter region. *Italian Journal of Animal Science*, 2(1), 67-70.

Griinari, J. M., & Bauman, D. E. (1999). Biosynthesis of conjugated linoleic acid and its incorporation into meat and milk in ruminants. *Advances in Conjugated Linoleic Acid Research*, 1(1), 180-200.

Grosclaude, F., Marie-Françoise, M., Brignon, G., Di Stasio, L., & Jeunet, R. (1987). A Mendelian polymorphism underlying quantitative variations of goat αs1-casein. *Génétique Sélection évolution*, 19(4), 399-412.

Guesnet, P., Antoine, J. M., de Lempdes Rochette, J. B., Galent, A., & Durand, G. (1993). Polyunsaturated fatty acid composition of human milk in France: changes during the course of lactation and regional differences. *European Journal of Clinical Nutrition*, 47(10), 700-710.

Gulati, S. K., Kitessa, S. M., Ashes, J. R., Fleck, E., Byers, E. B., Byers, Y. G., & Scott, T. W. (2000). Protection of conjugated linoleic acids from ruminal hydrogenation and their incorporation into milk fat. *Animal Feed Science and Technology*, 86(3-4), 139-148.

Guo, M. R. (2003). Goat's milk. In: Caballero, B., Trugo, L., Finglas, P. (Ed.), *Encyclopedia of Food Sciences and Nutrition*. Academic Press, London, UK, 2944–2949.

Guo, M. R., Dixon, P. H., Park, Y. W., Gilmore, J. A., & Kindstedt, P. S. (2001). Seasonal changes in the chemical composition of commingled goat milk. *Journal of Dairy Science*, 84, 79-83.

Ha, J. K., & Lindsay, R. C. (1991). Contributions of cow, sheep, and goat milks to characterizing branched-chain fatty acid and phenolic flavors in varietal Cheeses1. *Journal of Dairy Science*, 74(10), 3267-3274.

Haenlein, G. F. W., & Caccese, R. (1984). Goat milk versus cow milk. In: Haenlein, G. F. W., Ace, D. L. (Ed.), *Extension Goat Handbook*. USDA Publ., Washington, DC, U.S.A.

Haenlein, G. F., & Wendorff, W. L. (2006). *Sheep milk. Handbook of milk of non-bovine mammals,* 137-194.

Haenlein, G. F. W. (2004). Goat milk in human nutrition. *Small ruminant research*, 51(2), 155-163.

Hambraeus, L. (1982). Nutritional Aspects of Milk Proteins. In: Fox, P. F. (Ed.), Developments in Dairy Chemistry-1: Proteins. *Applied Science,* London, 289-313.

Hayes, H. C., & Petit, E. J. (1993). Mapping of the β-lactoglobulin gene and of an immunoglobulin M heavy chain-like sequence to homoeologous cattle, sheep, and goat chromosomes. *Mammalian Genome*, 4(4), 207-210.

Hayssen, V., & Blackburn, D. G. (1985). α-Lactalbumin and the origins of lactation. *Evolution,* 39(5), 1147-1149.

Horáčková, Š., Sedláčková, P., Sluková, M., & Plocková, M. (2015). The influence of whey, whey component and malt on the growth and acids production of lactobacilli in milk. *Czech Journal of Food Sciences*, 32(6), 526-531.

Ingham, B., Smialowska, A., Kirby, N. M., Wang, C., & Carr, A. J. (2018). A structural comparison of casein micelles in cow, goat and sheep milk using X-ray scattering. *Soft Matter,* 14(17), 3336-3343.

Inglingstad, R. A., Devold, T. G., Eriksen, E. K., Holm, H., Jacobsen, M., Liland, K. H., & Vegarud, G. E. (2010). Comparison of the digestion of caseins and whey proteins in equine, bovine, caprine and human milks by human gastrointestinal enzymes. *Dairy Science & Technology,* 90(5), 549-563.

Iverson, J. L., & Sheppard, A. J. (1989). Detection of adulteration in cow, goat, and sheep cheeses utilizing gas-liquid chromatographic fatty acid data. *Journal of Dairy Science*, 72(7), 1707-1712.

Jahreis, G., Fritsche, J., Möckel, P., Schöne, F., Möller, U., & Steinhart, H. (1999). The potential anticarcinogenic conjugated linoleic acid, cis-9, trans-11 C18: 2, in milk of different species: cow, goat, ewe, sow, mare, woman. *Nutrition Research*, 19(10), 1541-1549.

Jennes, R., & Parkash, S. (1971). Lack of a globule clustering agent in goat's milk. *Journal of Dairy Science*, 54(1), 123-126.

Jenness, R., & Patton, S. (1974). Physical properties of milk. *Fundamentals of Dairy Chemistry,* 402-441.

Jenness, R. (1982). Inter-species comparison of milk proteins. In: Fox, P.F. (Ed.), Developments in Dairy Chemistry-1: Proteins. *Applied Science*, New York, NY, U.A.S, 87–114.

Jenness, R. (1980). Composition and characteristics of goat milk: review 1968– 1979. *Journal of Dairy Science*, 63(10), 1605-1630.

Johansson, S. (2011). Goat Milk-Nutrition and health aspects. *Clinical Nutrition,* 25(3), 477-488.

Kuhn, N. J. (1983). The biosynthesis of lactose. In *'Biochemistry of Lactation'*. (Ed. TB Mepham.). Elsevier: Amsterdam. 159-76.

Kumar, A., Rout, P. K., & Roy, R. (2006). Polymorphism of β-lacto globulin gene in Indian goats and its effect on milk yield. *Journal of Applied Genetics*, 47(1), 49-53.

Kunz, C., Rudloff, S., Baier, W., Klein, N., & Strobel, S. (2000). Oligosaccharides in human milk: structural, functional, and metabolic aspects. *Annual Review of Nutrition*, 20(1), 699-722.

Lad, S. S., Aparnathi, K. D., Mehta, B., & Velpula, S. (2017). Goat milk in human nutrition and health-A review. *International Journal of Current Microbiology and Applied Sciences*, 6(5), 1781-1792.

Lara-Villoslada, F., Olivares, M., Jiménez, J., Boza, J., & Xaus, J. (2004). Goat milk is less immunogenic than cow milk in a murine model of atopy. *Journal of Pediatric Gastroenterology and Nutrition*, 39(4), 354-360.

Larson, B. L., & Smith, V. R. (1974). Lactation. A comprehensive treatise. Vol. II. *Biosynthesis and Secretion of Milk/Diseases*. Academic Press, New York, NY, U.S.A.

Lavigne, C., Zee, J. A., Simard, R. E., & Beliveau, B. (1989). Effect of Processing and Storage Conditions on the Fate of Vitamins B1, B2, and

C and on the Shelf-Life of Goat's Milk. *Journal of Food Science,* 54(1), 30-34.

LeDoux, M., Rouzeau, A., Bas, P., & Sauvant, D. (2002). Occurrence of trans-C18: 1 fatty acid isomers in goat milk: effect of two dietary regimens. *Journal of Dairy Science*, 85(1), 190-197.

Lee, K. W., Lee, H. J., Cho, H. Y., & Kim, Y. J. (2005). Role of the conjugated linoleic acid in the prevention of cancer. *Critical Reviews in Food Science and Nutrition*, 45(2), 135-144.

Loch, J. I., Bonarek, P., Polit, A., Świątek, S., Czub, M., Ludwikowska, M., & Lewiński, K. (2015). Conformational variability of goat β-lactoglobulin: Crystallographic and thermodynamic studies. *International Journal of Biological Macromolecules*, 72, 1283-1291.

Lönnerdal, B., & Lien, E. L. (2003). Nutritional and physiologic significance of α-lactalbumin in infants. *Nutrition Reviews*, 61(9), 295-305.

López-Aliaga, I., Díaz-Castro, J., Alférez, M. J. M., Barrionuevo, M., & Campos, M. S. (2010). A review of the nutritional and health aspects of goat milk in cases of intestinal resection. *Dairy Science & Technology,* 90(6), 611-622.

Luna, P., Fontecha, J., Juárez, M., & de la Fuente, M. A. (2005). Changes in the milk and cheese fat composition of ewes fed commercial supplements containing linseed with special reference to the CLA content and isomer composition. *Lipids*, 40(5), 445-454.

Ma, Y., Zhang, L., Wu, Y., & Zhou, P. (2019). Changes in milk fat globule membrane proteome after pasteurization in human, bovine and caprine species. *Food Chemistry*, 279, 209-215.

Mann, E. J. (1988). Ewes' and goats' milk and products, pt. 1[literature review]. *Dairy Industries International*, 53, 11-12.

Marchalonis, J. J., & Weltman, J. K. (1971). Relatedness among proteins: a new method of estimation and its application to immunoglobulins. *Comparative Biochemistry and Physiology Part B: Comparative Biochemistry*, 38(3), 609-625.

Markiewicz-Kęszycka, M., Czyżak-Runowska, G., Lipińska, P., & Wójtowski, J. (2013). Fatty acid profile of milk-a review. *Bulletin of the Veterinary Institute in Pulawy*, 57(2), 135-139.

Martin, P., Brignon, G., Furet, J. P., & Leroux, C. (1996). The gene encoding α_{s1}-casein is expressed in human mammary epithelial cells during lactation. *Le Lait*, *76*(6), 523-535.

Massart-Leen, A. M., De Pooter, H., Decloedt, M., & Schamp, N. (1981). Composition and variability of the branched-chain fatty acid fraction in the milk of goats and cows. *Lipids,* 16(5), 286-292.

McClenathan, D. T., & Walker, W. A. (1982). Food allergy: Cow milk and other common culprits. *Postgraduate Medicine*, 72(5), 233-239.

Mercier, J. C., & Chobert, J. M. (1976). Comparative study of the amino acid sequences of the caseinomacropeptides from seven species. *FEBS letters,* 72(2), 208-214.

Michalski, M. C., Gassi, J. Y., Famelart, M. H., Leconte, N., Camier, B., Michel, F., & Briard, V. (2003). The size of native milk fat globules affects physico-chemical and sensory properties of Camembert cheese. *Le lait,* 83(2), 131-143.

Moatsou, G., Hatzinaki, A., Samolada, M., & Anifantakis, E. (2005). Major whey proteins in ovine and caprine acid wheys from indigenous greek breeds. *International Dairy Journal*, 15(2), 123-131.

Mwenze, P. M. (2015). Functional properties of goats' milk: A review. *Research Journal of Agriculture and Environmental Management*, 4(9), 343-349.

Oftedal, O. T. (1984). Lactation in the dog: milk composition and intake by puppies. *The Journal of Nutrition*, 114(5), 803-812.

Oftedal, O. T., Iverson, S. J., & Jensen, R. G. (1995). *Handbook of milk composition*. Sydney Australia.

Oftedal, O. T., & Jenness, R. (1988). Interspecies variation in milk composition among horses, zebras and asses (Perissodactyla: Equidae). *Journal of Dairy Research*, 55(1), 57-66.

Park, Y. W., & Chukwu, H. I. (1988). Macro-mineral concentrations in milk of two goat breeds at different stages of lactation. *Small Ruminant Research*, 1(2), 157-166.

Park, Y. W., Juárez, M., Ramos, M., & Haenlein, G. F. W. (2007). Physico-chemical characteristics of goat and sheep milk. *Small Ruminant Research,* 68(1-2), 88-113.

Park, Y. W., & Haenlein, G. F. W. (2008). *Handbook of milk of non-bovine mammals*. sl. John Wiley & Sons, Hoboken NJ, U.S.A.

Park, Y. W. (1991). Relative buffering capacity of goat milk, cow milk, soy-based infant formulas, and commercial nonprescription antacid drugs. *Journal of Dairy Science*. 74(10), 3326-3333.

Parkash, S., & Jenness, R. (1968). The composition and characteristics of goat's milk: A review. *Dairy Sci. Abstr*. 30(2), 67-87.

Pearse, M. J., Linklater, P. M., Hall, R. J., & Mackinlay, A. G. (1986). Effect of casein micelle composition and casein dephosphorylation on coagulation and syneresis. *Journal of Dairy Research*, 53(3), 381-390.

Pena, R. N., Sánchez, A., & Folch, J. M. (2000). Characterization of genetic polymorphism in the goat β-lactoglobulin gene. *Journal of Dairy Research*, 67(2), 217-224.

Pérez, M. D., & Calvo, M. (1995). Interaction of β-lactoglobulin with retinol and fatty acids and its role as a possible biological function for this protein: a review. *Journal of Dairy Science*, 78(5), 978-988.

Pessler, F., & Nejat, M. (2004). Anaphylactic reaction to goat's milk in a cow's milk-allergic infant. *Pediatric Allergy and Immunology*, 15(2), 183-185.

Precht, D. (1995). Variation of trans fatty acids in milk fats. *Zeitschrift für Ernährungswissenschaft,* 34(1), 27-29.

Prieels, J. P., Poortmans, J., Dolmans, M., & Leonis, J. (1975). Immunological cross-reactions of α-Lactalbumins from different species. *European Journal of Biochemistry*, 50(3), 523-527.

Puri, B. R., & Parkash, S. (1962). Studies on physico-chemical properties of milk. XII. The renneting time of milk of different species. *Indian Journal of Dairy Science*, 15, 114-122.

Recio, I., Pérez-Rodríguez, M. L., Amigo, L., & Ramos, M. (1997). Study of the polymorphism of caprine milk caseins by capillary electrophoresis. *Journal of Dairy Research*, 64(4), 515-523.

Remeuf, F. (1992). Physico-chemical properties of goat milk in relation to processing characteristics. In: *Proceedings of the National Symposium on Dairy Goat Production and Marketing*, Oklahoma City, OK, 98-110.

Renner, E., Schaafsma, G., & Scott, K. J. (1989). Micronutrients in Milk, In: Renner, E. (Ed.), *Micronutrients in Milk and Milk-based Products*. Elsevier Applied Science, New York, U.S.A, 1–70.

Rincón, F., Moreno, R., Zurera, G., & Amaro, M. (1994). Mineral composition as a characteristic for the identification of animal origin of raw milk. *Journal of Dairy Research*, 61(1), 151-154.

Robinson, F. (2001). Goats milk–a suitable hypoallergenic alternative? *British Food Journal,* 103(3), 198-208.

Rodríguez-Alcalá, L. M., Harte, F., & Fontecha, J. (2009). Fatty acid profile and CLA isomers content of cow, ewe and goat milks processed by high pressure homogenization. *Innovative Food Science & Emerging Technologies,* 10(1), 32-36.

Selner, D. R., & Schultz, L. H. (1980). Effects of feeding oleic acid or hydrogenated vegetable oils to lactating cows. *Journal of Dairy Science*, 63(8), 1235-1241.

Selvaggi, M., Laudadio, V., Dario, C., & Tufarelli, V. (2014). Major proteins in goat milk: an updated overview on genetic variability. *Molecular Biology Reports*, 41(2), 1035-1048.

Sindayikengera, S., & Xia, W. S. (2006). Nutritional evaluation of caseins and whey proteins and their hydrolysates from Protamex. *Journal of Zhejiang University Science* B, 7(2), 90-98.

Slačanac, V., Božanić, R., Hardi, J., Rezessyné Szabó, J. U. D. I. T., Lučan, M., & Krstanović, V. (2010). Nutritional and therapeutic value of fermented caprine milk. *International Journal of Dairy Technology*, 63(2), 171-189.

Spuergin, P., Walter, M., Schiltz, E., Deichmann, K., Forster, J., & Mueller, H. (1997). Allergenicity of α-caseins from cow, sheep, and goat. *Allergy,* 52(3), 293-298.

Stanton, C., Murphy, J., McGrath, E., & Devery, R. (2003). Animal feeding strategies for conjugated linoleic acid enrichment of milk. *Advances in Conjugated Linoleic Acid Research* 129-151.

Stewart, A. F., Bonsing, J., Beattie, C. W., Shah, F., Willis, I. M., & Mackinlay, A. G. (1987). Complete nucleotide sequences of bovine alpha S2-and beta-casein cDNAs: comparisons with related sequences in other species. *Molecular Biology and Evolution*, 4(3), 231-241.

Strzałkowska, N., Jóźwik, A., Bagnicka, E., Krzyżewski, J., Horbańczuk, K., Pyzel, B., & Horbańczuk, J. O. (2009). Chemical composition, physical traits and fatty acid profile of goat milk as related to the stage of lactation. *Animal Science Papers and Reports*, 27(4), 311-320.

Talpur, F. N., Bhanger, M. I., & Memon, N. N. (2009). Milk fatty acid composition of indigenous goat and ewe breeds from Sindh, Pakistan. *Journal of Food Composition and Analysis*, 22(1), 59-64.

Tomotake, H., Okuyama, R., Katagiri, M., Fuzita, M., Yamato, M., & Ota, F. (2006). Comparison between Holstein cow's milk and Japanese-Saanen goat's milk in fatty acid composition, lipid digestibility and protein profile. *Bioscience, Biotechnology, and Biochemistry*, *70*(11), 2771-2774.

Tripaldi, C., Martillotti, F., & Terramoccia, S. (1998). Content of taurine and other free amino acids in milk of goats bred in Italy. *Small Ruminant Research*, 30(2), 127-136.

Tsakalidou, E., & Papadimitriou, K. (Eds.). (2016). *Non-bovine milk and milk products*. Academic Press, Elsevier Inc. Cambridge, MA, U.S.A.

Tsiplakou, E., & Zervas, G. (2008). Comparative study between sheep and goats on rumenic acid and vaccenic acid in milk fat under the same dietary treatments. *Livestock Science*, 119(1-3), 87-94.

Tziboula-Clarke, A. (2003). Goat milk. *Encyclopedia of Dairy Sciences*, 2, 1270–1279.

Tziboula, A., & Horne, D. S. (1999). The role of αs1-casein in the structure of caprine casein micelles. *International Dairy Journal*, 9(3-6), 173-178.

Walstra, P. (1999). *Dairy technology: principles of milk properties and processes*. Marcel Dekker, New York, NY, U.S.A.

Walstra, P., & Jenness, R. (1984). *Dairy chemistry & physics*. John Wiley & Sons, New York, NY, U.S.A.

Whitney, R. M., Brunner, J. R., Ebner, K. E., Farrell Jr, H. M., Josephson, R. V., Morr, C. V., & Swaisgood, H. E. (1976). Nomenclature of the proteins of cow's milk: Fourth revision. *Journal of Dairy Science*, 59(5), 795-815.

Williams, C. M. (2000). Dietary fatty acids and human health. *Annales de zootechnie,* 49(3), 165-180.

Yahyaoui, M. H., Pena, R. N., Sanchez, A., & Folch, J. M. (2000). Rapid communication: polymorphism in the goat beta-lactoglobulin proximal promoter region. *Journal of Animal Science-Menasha Then Albany Then Champaign Illinois*, 78(4), 1100-1101.

Žan, M., Stibilj, V., & Rogelj, I. (2006). Milk fatty acid composition of goats grazing on alpine pasture. *Small Ruminant Research*, 64(1-2), 45-52.

In: Goat Milk Chemistry …
Editor: Mingruo Guo
ISBN: 978-1-53617-299-7

Chapter 4

GOAT MILK PROTEOMICS

***Yuxue Sun*[1], *Cuina Wang*[1] *and Mingruo Guo*[2,3,*]**
[1]Department of Food Science and Engineering,
Jilin University, Changchun, Jilin, China
[2]Key Laboratory of Dairy Science, Northeast Agricultural University,
Ministry of Education, Harbin, Heilongjiang, China
[3]Department of Nutrition and Food Sciences,
College of Agriculture and Life Sciences, University of Vermont,
Burlington, Vermont, US

ABSTRACT

Goat milk has been widely used in the dairy industry due to its importance to health. Milk protein, as a key factor for milk quality, has always been the focus of studies. With the continuous development of analytical techniques, proteomics is increasingly applied for the analysis of milk proteins as an emerging technology, and more and more proteins are identified in milk. In this chapter, the major proteomics techniques commonly used in study of milk proteins including those in procedures of protein separation and characterization were described. Bioinformatics

* Corresponding Author's Email: mguo@uvm.edu.

analysis tools for analyzing the raw data of mass spectrometry, such as corresponding database, Gene Ontology (GO) annotations, Kyoto Gene and Genomic Encyclopedia (KEGG) pathways and protein-protein interaction (PPI) network analysis, etc., were also mentioned. The applications of proteomics technology in study of goat milk proteins especially in whey proteins and milk fat globule membrane (MFGM) proteins were summarized. The protein digestion properties of goat milk by proteomics method was also discussed in the last part of this chapter.

Keywords: goat milk, proteomics, whey protein, MFGM protein

INTRODUCTION

The term "proteome" was first published in the journal "Electrophoresis" by Wasinger et al., in the 1990s. Proteome was defined as all proteins encoded by the genome (Wasinger et al., 1995). Proteomics, as the name implies, refers to the discipline of studying proteome, and sometimes is also known as post-genomics. There are several definitions of proteomics currently. One is that proteomics is the study on a range of protein properties, including qualitative and quantitative analyses of proteins, differentially expression analysis, glycosylation analysis, phosphorylation analysis, and other post-translational modification analysis of proteins, by which various physiological and pathological biological processes are reflected at the protein level. In another one, proteins separation techniques are also included besides identification and characterization of proteins defined above.

Milk is the food for newborn mammals and also an important protein source for adults. Goat milk, as one of the major milk supplies in the dairy industry, is rich in nutrients and bioactivities. Besides providing nutritional functions, goat milk also plays an important role in the growth and development including immune system for infants and young children (Blum & Hammon, 2000; Haenlein, 2004; Moreno-Indias et al., 2012). Due to its excellent nutritional and functional properties, goat milk has been used as a substitute for breast milk (Maathuis, Havenaar, He, & Bellmann, 2017; Xu et al., 2015), and more and more consumers are choosing goat milk instead of cow milk. In recent years, increasing studies have explored milk proteins, especially some low-abundance proteins using proteomic techniques. Differently from those of traditional biochemical analysis methods, study on milk protein using proteomic techniques enables simultaneous, high resolution analysis of proteins in complex ones. Proteomics technology method has a wide range of characterizations that can achieve simple protein identifications as well as complex characterizations such as post-translational modifications and protein-protein interactions. Proteomics (in a broad sense) mainly includes two techniques: separation and characterization.

PROTEOMICS TECHNIQUES

In order to understand the more detailed protein composition and functional properties, the accuracy of the detected results is usually improved by two aspects: developing a higher sensitivity characterization technique on the one hand, and simplifying the target proteome on the other hand, that is, adding the processes of protein separation and purification before characterization to reduce the effect of non-target proteins on low-abundance milk proteins.

Separation Techniques

Protein separation is the first step in proteomics study and also the key to affecting the results of protein characterization. So far, there are two methods that most commonly used for separating proteins: electrophoretic separation and liquid phase separation.

Sodium Dodecyl Sulfate-Polyacrylamide Gel Electrophoresis (SDS-PAGE) is one of the most well-known protein separation technologies, which is often used in proteomics study in combination with other separation techniques to achieve more efficient separation.

Two-Dimensional Gel Electrophoresis (2-DE) is an electrophoresis technique based on SDS-PAGE combined with isoelectric focusing (IEF). 2-DE can distinguish proteins by two protein properties, molecular weight and isoelectric point. It is a very powerful tool for separating similar proteins and has been reported in many studies (Holt & Zeece, 1988; Westbrook, Yan, Wait, Welson, & Dunn, 2001).

Since the separation principle of 2-DE is based on the different isoelectric point and molecular weight of proteins, those proteins with similar isoelectric points, extremely high or low molecular weight and low abundance generally cannot achieve effective separation on the gel. The steps of protein pre-separation are usually added in advance to enhance separation satisfaction when using this technique.

Liquid phase separation is a more commonly used separation method in proteomics study. Among which, High Performance Liquid Chromatography (HPLC) has become an essential method for separating proteins and peptides currently. Compared with 2-DE, liquid phase separation has a higher sensitivity and wider dynamic detection range, and is easier to achieve automation, which makes up for the deficiency of 2-DE in protein separation process.

Additionally, liquid phase techniques with different separation mechanisms can be stacked in series to increase separation peak capacity and resolution, thereby meeting the effective separation of more complex samples. For example, two separations of the protein/peptide can be achieved by superposition Ion Exchange Chromatography and Reversed Phase Chromatography. This technique was successfully used in the study of yeast proteome in 2001 (Washburn, Wolters, & Yates, 2001). Two-dimensional, three-dimensional and multi-dimensional liquid chromatography techniques (i.e., superimposing two, three or more kinds of liquid chromatography) have been widely used in proteomics research currently. It is also possible to add a step of pre-separation before liquid phase separation to obtain a further satisfactory separation.

Characterization Techniques

The separation of protein/peptides is followed by characterization. Mass Spectrometry (MS) is a commonly used characterization technique in proteomics study. Through which, comprehensively characterization of complex sample proteins is achieved. MS can be divided into different categories based on their ionization source and mass analyzer components. There are several commonly used ionization sources such as matrix assisted laser desorption ionization (MALDI) ion sources and electrospray ionization (ESI) ion sources. With the development of technology, nano-electrospray ionization source (nESI) has also been applied to the proteomics studies. The combination use of nESI in series with nanoliter Liquid Chromatography allows for a more detailed analysis of protein/peptides by greatly controlling the flow rate of the sample. The mass analyzer includes the well-known quadrupole (Q), time-of-flight (TOF) analyzer, Fourier transform ion cyclotron resonance (FT-ICR) analyzer, ion trap (IT) and Orbitrap, etc. These mass analyzers also can be used alone or in series in studies for a better characterization.

Bioinformatics Analysis

The raw data obtained by mass spectrometry analysis is then matched with the corresponding database to obtain detail information of the protein/peptides. These databases are derived from the Uniprot database (https://www.uniprot.org/), National Center for Biotechnology Information (NCBI) database etc... For those identified/quantified protein, several bioinformatics analyses including Gene Ontology (GO) annotation, Kyoto Encyclopedia of Genes and Genomes (KEGG) pathway, cluster heat map, principal component analysis (PCA) and protein-protein interaction (PPI) network analysis can be performed. GO annotation is a tool for collecting and counting protein annotation terms at the functional level, by which, proteins can be categorized based on biological processes, cellular component, and molecular function (Ashburner et al., 2000; Gene Ontology,

2006). The KEGG pathway is a basic database that links genomic/proteomic information to its high-level functional pathway information (Kanehisa & Goto, 2000). These two analyses described above are often performed through the online Database for Annotation, Visualization and Integrated Discovery (DAVID) database and the KEGG online database. The cluster heat maps can be used to analyze the similarity between different protein sources, and classify different protein sources into different clusters, which can be obtained by various software such as ggplot2, Multiple Experiment Viewer (MeV) and Heml1.0. The Search Tool for the Retrieval of Interacting Genes (STRING) can be used to predict the PPI network, and then the PPI network can be optimized and analyzed by Cytoscape software.

IDENTIFICATION OF GOAT MILK PROTEINS

Milk is rich in nutrients such as protein, fat, lactose, vitamins and minerals etc., which can participate in the development of the body's bones, brain, metabolism and immune system (Blum & Hammon, 2000; Elizabeth, 2012; Moreno-Indias et al., 2012). Among these components, milk protein has a wide range of biological functions and is one of the main indicators for milk quality. Goat milk proteins consist of 70% casein, 25% whey protein and 5% milk fat globule membrane (MFGM) protein (Chen et al., 2019).

Casein

Casein is the main component of milk protein with a molecular weight of 20-25 kDa, which contains almost all the essential amino acids required by the body. In addition to providing nutrition, many bioactive peptides of casein are also important regulators of physiological activities. The protein composition of goat milk differs between different goat breeds and may present a regional trend. Costa et al., (Costa et al., 2014) studied the protein composition of Alpine and Saanen goat milk, results suggested that no αs_1-casein were observed in both milk and a higher amount of casein were observed in Saanen goat milk compared with that of Alpine goat. While in another study, caseins from two kinds of genotyped goat milk were identified using 2-DE combined with MALDI-TOF MS. The results showed that the total casein content in goat milk was positively correlated with αs_1-casein content and the total casein content of milk with strong alleles was much higher than that without strong alleles (Roncada et al., 2002).

Phosphorylation is one of the major modifications of casein and is widely used in the food industry. Eight phosphopeptides with 18 different phosphorylation sites were reported in casein phosphoproteome of goat milk (Olumee-Shabon & Boehmer, 2013). Among which, a tetra-phosphopeptide (N62−K86) was identified as a potential marker for milk quality.

Casein is a thoroughly studied protein in the study of milk-derived bioactive peptides currently, therefore, there are few studies focused on casein by proteomics methods, even more so regarding casein chemistry of goat milk. In recent years, the analysis of milk proteins by proteomics methods has mainly focused on the MFGM and whey proteome. Interestingly, caseins were present in MFGM as well as whey proteins (Anagnostopoulos et al., 2016; D'Ambrosio et al., 2008; Liao, Alvarado, Phinney, & Lonnerdal, 2011).

Whey Proteins

Although the proportion is as high as casein, whey protein is known as a protein with a wide range of biological activities (Chaneton, Sáez, & Bussmann, 2011; Saint-Sauveur, Gauthier, Boutin, & Montoni, 2008; Schmelzle et al., 2003).

In recent years, increasing studies have explored whey proteins of goat milk using proteomic techniques (Table 1), including characterization of whey proteome in many different goat breeds, comparison of whey proteome between goat and other species, and comparison of whey peoteome at different lactation stages, etc.

Table 1. Detailed studies on whey proteins in goat milk

Goat breeds	Number of animals sampled	Method	Number of identified proteins	Reference
Capra prisca goat (scattered across Greek)	17	SDS-PAGE, LC-MS/MS	595	(Anagnostopoulos et al., 2016)
Skopelos goat (scattered across Greek)	15	SDS-PAGE, LC-MS/MS	486	(Anagnostopoulos et al., 2016)
Camosciata goat (Sicilian, Italy)	100	CPLL, SDS-PAGE, LC-MS/MS	452	(Cunsolo et al., 2015)
Guanzhong goat (Jilin, China)	30	LC-MS/MS	283	(Sun et al., 2019a)
Commercial tank milk (Beijing, China)	-	LC-MS/MS	213	(Lu et al., 2018)
Majorera goat (Canary Islands, Spain)	6	2-DE, LC-MS/MS	-	(Hernandez-Castellano et al., 2016)
Xinong Saanen goat (Shanxi, China)	30	LC-MS/MS	643	(Sun et al., 2019c)

Anagnostopoulos et al., (2016) identified 486 and 595 proteins in whey fractions from two milk samples of indigenous Greek goats, respectively, using SDS-PAGE combined with LC-MS/MS. β-Lactoglobulin, mostly known as an allergen of cow milk, was observed in both goat milk, and a higher content of β-lactoglobulin was shown in Capra prisca (one of the indigenous Greek goat breeds of this study) milk. For "Camosciata" goat breed, a total of 452 whey proteins were found in milk by coupling combinatorial peptide ligand library (CPLL) technology, SDS-PAGE and nano LC-MS/MS. Most of these proteins fell into biological process of metabolic process, cellular component of extracellular region and molecular function of catalytic activity (Cunsolo et al., 2015).

Moreover, Cunsolo et al., (2015) also compared whey proteins of "Camosciata" goat milk with those of bovine milk and revealed that approximately 45% of proteins were shared in both, including several known allergens and defense/immunity-related proteins. We previously performed a comparison of the whey proteins between Guanzhong goat and Holstein cow milk using proteomics technology, and 283, 159 whey proteins were identified in goat and cow milk, respectively. These two kinds of whey proteins enriched into different biological processes and molecular functions while both fell into cellular component of extracellular exosome (Sun, Wang, Sun, & Guo, 2019a).

The comparison analysis of whey proteins between goat milk and other species were also reported previously. A total of 198, 169, 213 and 128 proteins were reported found in serum of human, cow, goat and yak milk, and high level of complement proteins and acute phase proteins were observed in goat milk serum (Lu et al., 2018).

The lactation stage may be a factor influencing goat milk, and several previous studies have focused on whey proteomes of goat milk at different lactation stages (Hernandez-Castellano, Almeida, Castro, & Arguello, 2014; Hernandez-Castellano, Almeida, Renaut, Arguello, & Castro, 2016). Hernandez-Castellano et al., (2016) performed the proteomics analysis of colostrum and milk from three ruminant species of Holstein cow, Canarian sheep and Majorera goats. Although the concentrations of immunoglobulins in colostrum and milk of the three species were similar, differences in several immune components can be detected using proteomics methods. We previously characarized and compared the whey proteins in colostrum and mature milk of Xinong Saanen goat. During which, whey proteins of goat colostrum showed differences in functional categories, GO annotation, KEGG pathway and PPI network. Immune-related proteins were abundant in both milk. More disease-related proteins were found in colostrum, while mature milk showed more metabolism-related proteins (Sun, Wang, Sun, & Guo, 2019c).

Milk Fat Globule Membrane Proteins

The milk fat globule membrane (MFGM) is a three-layer ordered membrane with a thickness of 10-20 nm. MFGM proteins are very low in content, about 5% of total protein in goat milk, which has significantly different composition and functions from those of whey proteins (Singh, 2006). These MFGM proteins were reported to be associated with many diseases such as multiple sclerosis, autism and coronary heart disease (Riccio, 2004).

Studies of MFGM proteins have been the focus of milk proteins for decades. More and more MFGM proteins of goat milk have been identified with the development of proteomics techniques, and the recent studies are listed in Table 2. The major proteins of goat milk such as fatty acid synthase, xanthine oxidase, butyrophilin, lactadherin (also known as milk fat globule-EGF 8), and adipophilin were all identified using SDS-PAGE coupled with MALDI-TOF MS.

Among which, the structure of O-glycans was found existing in fatty acid synthase and xanthine oxidase (Cebo, Caillat, Bouvier, & Martin, 2010). In another study, 207 protein spots were separated from caprine MFGM by 2-DE. Besides the major caprine MFGM proteins, various protein isoforms of post-translational modifications were observed on the 2-DE map such as modified butyrophilin, lactadherin, perilipin and proteose peptone-3.

Several caseins and whey proteins also detected in caprine MFGM proteins, which can be explained by the binding between these caseins/whey proteins and damaged MFGs (Spertino et al., 2012). There were 175 MFGM proteins observed from a commercial goat milk. Among them, stomatin and xanthine dehydrogenase/oxidase were significantly enriched in goat MFGM (Lu, Wang, et al., 2016). Our previous study reported 593 MFGM proteins of goat milk using nano LC-Orbitrap MS/MS, which were mostly involved in oxidation-reduction biological process and metabolism processes-related pathways (Sun et al., 2019a).

Table 2. Detailed studies on MFGM proteins in goat milk

Goat breeds	Number of animals sampled	Method	Number of identified proteins	Reference
Goat (Avord, France).	8	SDS-PAGE, LC-MS	-	(Cebo et al., 2010)
Goat (Piemonte, Italy)	-	SDS-PAGE, 2-DE, LC-MS/MS	207 spots from 2-DE	(Spertino et al., 2012)
Commercial tank milk (Beijing, China)	-	LC-MS/MS	175	(Lu, Wang, et al., 2016)
Guanzhong goat (Jilin, China)	30	LC-MS/MS	593	(Sun et al., 2019a)
Saanen goat (Jiangsu, China)	35	LC-MS/MS	137	(Ma, Zhang, Wu, & Zhou, 2019)
Guanzhong goat (Shanxi, China)	5	LC-MS/MS	423	(Lu, Liu, et al., 2016).

In addition to differences with cow milk (Cebo et al., 2010; Sun et al., 2019a; Zamora, Guamis, & Trujillo, 2009), many differences in MFGM proteins between goat and other species were reported. MFGM proteins in human, bovine, goat, and camel milk were mainly involved in biological process of protein transport, cellular component of membrane and molecular function of GTP binding, while these four kinds of MFGM proteins varied widely among species in lipids synthetic, secretory related proteins (such as xanthine dehydrogenase/oxidase) and host defense proteins (Lu, Wang, et al., 2016).

MFGM proteins of goat milk also showed significant differences at different lactation stages. A comparison of MFGM proteomes in Guanzhong goat colostrum and mature milk was previously studied, and 189 significantly different MFGM proteins were observed. One hundred and eighty-one of which were higher expressed in colostrum, especially several acute inflammatory response proteins. While mature milk showed only 8 higher abundant proteins compared with those of colostrum, including xanthine dehydrogenase/oxidase, stomatin and mucin, etc. (Lu, Liu, et al., 2016).

CHARACTERIZATION OF DIGESTION PROPERTIES IN GOAT MILK

The composition and content of goat milk proteins are affected by factors such as species, lactation stages and regions. And differences in the composition and content of milk proteins definitely give them different digestion properties. Thus, it is necessary to study the digestion properties of milk proteins by proteomics methods. A number of studies were focusing on the digestion properties of milk proteins including the study on the proteome of digested goat milk.

The single protein fraction of goat milk such as whey protein can be directly studied as the digested material. Caprine whey protein concentrate (81%, w/v) was collected and then digested using in vitro adult model with human gastrointestinal enzymes in a previous study. The hydrolysates after gastric digestion showed few peptides and less antibacterial effect compred with those after gastrointestinal digestion. At the end of the gastrointestinal digestion, a total of 106 peptides were identified in the hydrolysates, which were derived from β-lactoglobulin, β-casein derivatives of γ-caseins, κ-casein glycomacropeptide and lactoferrin. Twenty-two of these peptides were synthesized and tested for their antibacterial activity. Interestingly, all synthesized peptides were less active than those fractionated hydrolysates (Almaas et al., 2011).

The digestion properties of goat milk are also compared to those of cow milk. During a comparison of gastric digestion in goat and cow milk using in vitro infant model with commercial porcine pepsin, over 350 peptides were identified in hydrolysates of goat and cow milk. The majority of peptides were derived from casein regardless of the species, while goat milk showed more κ-casein-derived peptides and less αs_1-casein-derived peptides than cow milk. Minor milk proteins accounted for only 5% of peptides, which were composed of different proteins in goat milk and cow milk (Hodgkinson, Wallace, Smolenski, & Prosser, 2019).

Difference in digestion properties also occurs in milk at different lactation stages. We previously characterized the digestion properties of

Xinong Saanen goat colostrum and mature milk by simulating infant gastrointestinal digestion. After the whole gastrointestinal digestion, 30 and 42 unique peptides (molecular weight below 10 kDa) were identified in hydrolysates of colostrum and mature milk, respectively. Among them, the casein-derived peptides identified in colostrum and mature milk accounted for 76.67% and 59.53%, respectively. β-Casein was the most abundant in colostrum while that in mature milk was α_{s1}-casein. Several minor proteins such as enterotoxin-binding glycoprotein PP20K, butyrophilin subfamily 1 member A1 and perilipin were only detected in hydrolysates of mature milk. It seems that colostrum showed lower digestion degree, but it had more bioactive peptides (molecular weight < 10 kDa) than mature milk. The difference in bioactive peptides contained in both milk was mainly observed in αs_1-casein (Sun, Wang, Sun, & Guo, 2019b). And these bioactive peptides may play an important role in growth of newborns.

In these digestion studies of goat milk, fewer minor proteins were detected. One reason may be that many of these minor proteins are post-translational modified proteins, which may increase their resistance to hydrolysis. Another explanation is the low content of these proteins in goat milk. Although not much in quantity, these proteins have their own biological functions in health (Donovan, 2016; Liu & Newburg, 2013; Niklas, Erik, Olle, Bo, & Magnus, 2014; Smolenski et al., 2007). Procedures of pre-separation and enrichment may be used before analyses for a better characterization in future research.

CONCLUSION

With the development of proteomics technology, more proteins and their functional activities have been identified in goat milk. Goat milk is one of the potential alternative sources to cow milk. All these data may provide useful information about utilization of goat milk for goat milk-based products such as infant formula formulation.

References

Almaas, H., Eriksen, E., Sekse, C., Comi, I., Flengsrud, R., Holm, H., Jensen, E., Jacobsen, M., Langsrud, T., & Vegarud, G. E. (2011). Antibacterial peptides derived from caprine whey proteins, by digestion with human gastrointestinal juice. *British Journal of Nutrition*, 106(6), 896-905.

Anagnostopoulos, A. K., Katsafadou, A. I., Pierros, V., Kontopodis, E., Fthenakis, G. C., Arsenos, G., Karkabounas, S. C., Tzora, A., Skoufos, I., & Tsangaris, G. T. (2016). Milk of Greek sheep and goat breeds; characterization by means of proteomics. *Journal of Proteomics*, 147, 76-84.

Ashburner, M., Ball, C. A., Blake, J. A., Botstein, D., Butler, H., Cherry, J. M., Davis, A. P., Dolinski, K., Dwight, S. S., Eppig, J. T., Harris, M. A., Hill, D. P., Issel-Tarver, L., Kasarskis, A., Lewis, S., Matese, J. C., Richardson, J. E., Ringwald, M., Rubin, G. M., & Sherlock, G. (2000). Gene ontology: tool for the unification of biology. The Gene Ontology Consortium. *Gene*, 25(1), 25-29.

Blum, J. W., & Hammon, H. (2000). Colostrum effects on the gastrointestinal tract, and on nutritional, endocrine and metabolic parameters in neonatal calves. *Livestock Production Science*, 66(2), 151-159.

Cebo, C., Caillat, H., Bouvier, F., & Martin, P. (2010). Major proteins of the goat milk fat globule membrane. *Journal of Dairy Science*, 93(3), 868-876.

Chaneton, L., Sáez, J. P., & Bussmann, L. E. (2011). Antimicrobial activity of bovine β-lactoglobulin against mastitis-causing bacteria. *Journal of Dairy Science*, 94(1), 138-145.

Chen, D., Li, X., Zhao, X., Qin, Y., Wang, J., & Wang, C. (2019). Comparative proteomics of goat milk during heated processing. Fo*od Chemistry*, 275, 504-514.

Costa, W. K., Souza, E. L., Beltrao-Filho, E. M., Vasconcelos, G. K., Santi-Gadelha, T., de Almeida Gadelha, C. A., Franco, O. L., Rita de Cassia Ramos do Egypto, Queiroga., & Magnani, M. (2014). Comparative protein composition analysis of goat milk produced by the Alpine and Saanen breeds in northeastern Brazil and related antibacterial activities. *PLoS One*, 9(3), e93361.

Cunsolo, V., Fasoli, E., Saletti, R., Muccilli, V., Gallina, S., Righetti, P. G., & Foti, S. (2015). Zeus, Aesculapius, Amalthea and the proteome of goat milk. *Journal of Proteomics*, 128, 69-82.

D'Ambrosio, C., Arena, S., Salzano, A. M., Renzone, G., Ledda, L., & Scaloni, A. (2008). A proteomic characterization of water buffalo milk fractions describing PTM of major species and the identification of minor components involved in nutrient delivery and defense against pathogens. *Proteomics*, 8(17), 3657-3666.

Donovan, S. M. (2016). The Role of Lactoferrin in Gastrointestinal and Immune Development and Function: A Preclinical Perspective. *Journal of Pediatrics*, 173, S16-S28.

Elizabeth, M. G. (2012). Ruth goes home: an adult's use of human milk. *Breastfeed Rev*, 20(3), 44-48.

Gene Ontology, C. (2006). The Gene Ontology (GO) project in 2006. *Nucleic Acids Research*, 34(Database issue), D322-326.

Haenlein, G. F. W. (2004). Goat milk in human nutrition. *Small Ruminant Research*, 51(2), 155-163.

Hernandez-Castellano, L. E., Almeida, A. M., Castro, N., & Arguello, A. (2014). The colostrum proteome, ruminant nutrition and immunity: a review. *Current Protein & Peptide Science*, 15(1), 64-74.

Hernandez-Castellano, L. E., Almeida, A. M., Renaut, J., Arguello, A., & Castro, N. (2016). A proteomics study of colostrum and milk from the two major small ruminant dairy breeds from the Canary Islands: a bovine milk comparison perspective. *Journal of Dairy Research*, 83(3), 366-374.

Hodgkinson, A. J., Wallace, O. A. M., Smolenski, G., & Prosser, C. G. (2019). Gastric digestion of cow and goat milk: Peptides derived from simulated conditions of infant digestion. *Food Chemistry*, 276, 619-625.

Holt, D. L., & Zeece, M. G. (1988). Two-Dimensional Electrophoresis of Bovine Milk Proteins 1. *Journal of Dairy Science*, 71(8), 2044-2050.

Kanehisa, M., & Goto, S. (2000). KEGG: kyoto encyclopedia of genes and genomes. *Nucleic Acids Research*, 28(1), 27-30.

Liao, Y. L., Alvarado, R., Phinney, B., & Lonnerdal, B. (2011). Proteomic Characterization of Human Milk Fat Globule Membrane Proteins during a 12 Month Lactation Period. *Journal of Proteome Research*, 10(8), 3530-3541.

Liu, B., & Newburg, D. S. (2013). Human Milk Glycoproteins Protect Infants Against Human Pathogens. *Breastfeeding Medicine*, 8(4), 354-362.

Lu, J., Liu, L., Pang, X. Y., Zhang, S. W., Jia, Z. H., Ma, C. L., Zhao, L. L., & Lv, J. P. (2016). Comparative proteomics of milk fat globule membrane in goat colostrum and mature milk. *Food Chemistry*, 209, 10-16.

Lu, J., Wang, X. Y., Zhang, W. Q., Liu, L., Pang, X. Y., Zhang, S. W., & Lv, J. P. (2016). Comparative proteomics of milk fat globule membrane in different species reveals variations in lactation and nutrition. *Food Chemistry*, 196, 665-672.

Lu, J., Zhang, S. W., Liu, L., Pang, X. Y., Ma, C. L., Jiang, S. L., & Lv, J. P. (2018). Comparative proteomics analysis of human and ruminant milk serum reveals variation in protection and nutrition. *Food Chemistry*, 261, 274-282.

Ma, Y., Zhang, L. N., Wu, Y. Y., & Zhou, P. (2019). Changes in milk fat globule membrane proteome after pasteurization in human, bovine and caprine species. *Food Chemistry*, 279, 209-215.

Maathuis, A., Havenaar, R., He, T., & Bellmann, S. (2017). Protein Digestion and Quality of Goat and Cow Milk Infant Formula and Human Milk under Simulated Infant Conditions. *Journal of Pediatric Gastroenterology and Nutrition*, 65(6), 661-666.

Moreno-Indias, I., Sanchez-Macias, D., Castro, N., Morales-delaNuez, A., Hernandez-Castellano, L. E., Capote, J., & Arguello, A. (2012). Chemical composition and immune status of dairy goat colostrum fractions during the first 10 h after partum. *Small Ruminant Research*, 103(2-3), 220-224.

Niklas, T., Erik, D. F., Olle, H., Bo, L. N., & Magnus, D. F. (2014). Neurodevelopment, nutrition, and growth until 12 mo of age in infants fed a low-energy, low-protein formula supplemented with bovine milk fat globule membranes: a randomized controlled trial. *American Journal of Clinical Nutrition*, 99(4), 860-868.

Olumee-Shabon, Z., & Boehmer, J. L. (2013). Detection of Casein Phosphopeptides in Goat Milk. *Journal of Proteome Research*, 12(6), 3034-3041.

Riccio, P. (2004). The proteins of the milk fat globule membrane in the balance. *Trends in Food Science & Technology*, 15(9), 458-461.

Roncada, P., Gaviraghi, A., Liberatori, S., Canas, B., Bini, L., & Greppi, G. F. (2002). Identification of caseins in goat milk. *Proteomics*, 2(6), 723-726.

Saint-Sauveur, D., Gauthier, S. F., Boutin, Y., & Montoni, A. (2008). Immunomodulating properties of a whey protein isolate, its enzymatic digest and peptide fractions. *International Dairy Journal*, 18(3), 260-270.

Schmelzle, H., Wirth, S., Skopnik, H., Radke, M., Knol, J., Bockler, H. M., Bronstrup, A., Wells, J., & Fusch, C. (2003). Randomized double-blind study of the nutritional efficacy and bifidogenicity of a new infant formula containing partially hydrolyzed protein, a high beta-palmitic acid level, and nondigestible oligosaccharides. *Journal of Pediatric Gastroenterology and Nutrition*, 36(3), 343-351.

Singh, H. (2006). The milk fat globule membrane-A biophysical system for food applications. *Current Opinion in Colloid & Interface Science*, 11(2-3), 154-163.

Smolenski, G., Haines, S., Kwan, F. Y. S., Bond, J., Farr, V., Davis, S. R., Stelwagen, K., & Wheeler, T. T. (2007). Characterisation of host defence proteins in milk using a proteomic approach. *Journal of Proteome Research*, 6(1), 207-215.

Spertino, S., Cipriani, V., De Angelis, C., Giuffrida, M. G., Marsano, F., & Cavaletto, M. (2012). Proteome profile and biological activity of caprine, bovine and human milk fat globules. *Molecular Biosystems*, 8(4), 967-974.

Sun, Y. X., Wang, C. N., Sun, X. M., & Guo, M. R. (2019a). Comparative Proteomics of Whey and Milk Fat Globule Membrane Proteins of Guanzhong Goat and Holstein Cow Mature Milk. *Journal of Food Science,* 84(2), 244-253.

Sun, Y. X., Wang, C. N., Sun, X. M., & Guo, M. R. (2019b). Protein digestion properties of Xinong Saanen goat colostrum and mature milk using in vitro digestion model. *Journal of the Science of Food and Agriculture*, 99(13), 5819-5825.

Sun, Y. X., Wang, C. N., Sun, X. M., & Guo, M. R. (2019c). Proteomic analysis on whey proteins in colostrum and mature milk of Xinong Saanen goat. *Journal of Dairy Science*, In press.

Washburn, M. P., Wolters, D., & Yates, J. R. (2001). Large-scale analysis of the yeast proteome by multidimensional protein identification technology. *Nature Biotechnology*, 19(3), 242-247.

Wasinger, V. C., Cordwell, S. J., Cerpa-Poljak, A., Yan, J. X., Gooley, A. A., Wilkins, M. R., Duncan, M. W., Harris, R., Williams, K. L., & Humphery-Smith, I. (1995). Progress with gene-product mapping of the Mollicutes: Mycoplasma genitalium. *Electrophoresis*, 16(7), 1090-1094.

Westbrook, J. A., Yan, J. X., Wait, R., Welson, S. Y., & Dunn, M. J. (2001). Zooming-in on the proteome: Very narrow-range immobilised pH gradients reveal more protein species and isoforms. *Electrophoresis*, 22(14), 2865-2871.

Xu, M. H., Wang, Y. B., Dai, Z. Y., Zhang, Y. C., Li, Y., & Wang, J. B. (2015). Comparison of growth and nutritional status in infants receiving goat milk-based formula and cow milk-based formula: a randomized, double-blind study. *Food & Nutrition Research*, 59(1), 28613.

Zamora, A., Guamis, B., & Trujillo, A. J. (2009). Protein composition of caprine milk fat globule membrane. *Small Ruminant Research*, 82(2-3), 122-129.

In: Goat Milk Chemistry …
Editor: Mingruo Guo

ISBN: 978-1-53617-299-7

Chapter 5

GOAT MILK FLAVOR

***Cuina Wang*[1], *Ce Wang*[1] and *Mingruo Guo*[2,3,*]**
[1]Department of Food Science, College of Food Science and Engineering, Jilin University, Changchun, Jilin, China
[2]Department of Nutrition and Food Sciences, College of Agriculture and Life Sciences, University of Vermont, Burlington, Vermont, US
[3]Key Laboratory of Dairy Science (Northeast Agricultural University), Ministry of Education, Harbin, Heilongjiang, China

ABSTRACT

Goat milk and its dairy products have a strong unique goaty flavor. This flavor may pose a challenge to manufacturers and impact products acceptance for certain consumers. In this chapter, definition of "goaty flavor" was introduced and then possible origins for the flavor and approaches to minimize the flavor were summarized. Goaty flavor is a typical property of goat milk given by a complex sensation of taste together with odor or aroma. Short-medium chain free fatty acids (SM-FFA) and volatile branched-chain fatty acids (BCFA) may be responsible for this unique flavor. Addition of plant- or fruit-based ingredients, entrapping agent or lactic acid fermentation can reduce the flavor effectively. This

* Corresponding Author's Email: mguo@uvm.edu.cn.

chapter may provide useful information for those who are not in favor of the unique flavor of goat milk.

Keywords: goaty flavor, free fatty acids, plant or fruit-based ingredient, β-cyclodextrin, whey protein, probiotic

DEFINITION OF GOATY FLAVOR

Goat milk may have better digestibility and certain therapeutic values in medicine and human nutrition in comparison with cow milk (Park, Juárez, Ramos, & Haenlein, 2007). However, for certain consumers, goat milk and its dairy products present a lower overall sensory acceptance due to its unpleasant goaty flavor. Flavor is the most important quality of milk in general, especially for goat milk. Goaty flavor is described as a typical property of goat milk given by a complex sensation of taste (gustatory sense) together with odor or aroma (olfactory sense) (Wilson & Baietto, 2009). Goaty odor or aroma is the characteristic odor (aroma) of goat milk perceived by the sensory (olfactory) cells in the nasal cavity by air passing through the nose or the mouth and nasopharynx while several volatile goaty flavor molecules are inhaled (Pereira et al., 2010). Goat milk flavor is irritating, bitter and soapy (Skjevdal, 1979) and can be perceived by consumers, even in goat milk yogurt with added fruit pulp (Senaka Ranadheera, Evans, Adams, & Baines, 2012).

In terms of consumers' attitudes towards the goaty flavor, Park and Drake (2005) stated that goat milk should have a mild, neutral and appealing flavor, which includes acceptable, attractive odor and taste. However, Mowlem (2005) stated that goats have had very bad publicity for many years and as a result considerable prejudice against goat products sold in the United Kingdom (UK) and people who are not goat enthusiasts will describe the milk as "strong, smelly, salty or sweet" (Ribeiro & Ribeiro, 2010). During the manufacture of some varieties of cheese, a relatively strong "goat like" flavor is preferred, but for other products the absence of characterizing flavor is favorable (Ribeiro & Ribeiro, 2010).

Chemistry and Possible Origins of Goaty Flavor

Short-Medium Chain Free Fatty Acids (SM-FFA)

Table 1. Fatty acid profile (g/100 g milk) of goat and cow milk

	Goat milk	Cow milk
C4:0 butyric	0.13	0.11
C6:0 caproic	0.09	0.06
C8:0 capryic	0.10	0.04
C10:0 capric	0.26	0.08
C12:0 lauric	0.12	0.09
C14:0 myristic	0.32	0.34
C16:0 palmitic	0.91	0.88
C18:0 stearic	0.44	0.40
C6-14: total MCT	0.89	0.61
C4-18: total SAFA	2.67	2.08
C16:1: palmitoleic	0.08	0.08
C18:1: oleic	0.98	0.84
C16:1-22:1 total MUFA	1.11	0.96
C18:2 linoleic	0.11	0.08
C18:3 linoleic	0.04	0.05
C18:2-18:3 total PUFA	0.15	0.12

Note: MCT: medium chain triglycerides; SAFA: saturated fatty acids; MUFA: monounstaturated fatty acids; PUFA: polyounstaturated fatty acids. Reference: (Posati & Orr, 1976; Cousin, 1982).

Ruminant milk fats are composed of more than 400 fatty acids with chain length ranging from C4 to C28 (Alonso, Fontecha, Lozada, Fraga, & Juárez, 1999; Schroeder, Markus, Vetter, & Walter, 2013). Most of the fatty acids are in the form of triglycerides. Compared with cow milk, goat milk shows a lot of difference in composition and content of fatty acid (Table 1). Free fatty acids (FFA) are released from triglycerides hydrolyzed by endogenous lipoprotein lipase (LPL). Free fatty acids content of goat milk is 3.11 μeq/mL and that of cow milk is 3.0 μeq/mL (Agnihotri & Prasad,

1993). Among this, content of short-medium chain free fatty acids (SM-FFA) plays an important role in the unique flavor of goat milk (Chilliard, Ferlay, Rouel, & Lamberet, 2003). SM-FFA as caproic (C_6), caprylic (C_8), and capric (C_{10}) acids are more abundant in goat milk than cow milk (Costa et al., 2014). The presence of high content of C_6-C_8 in goat milk and dairy products may impart the characteristic unpleasant flavor (Chilliard et al., 2003; Chaminda Senaka Ranadheera, Naumovski, & Ajlouni, 2018). Flavor intensity increased in Italian cheese as SM-FFA concentrations increased (Ha and Lindsay, 1991). There is a strong positive correlation (r = 0.83) between the amount of SM-FFA and the goaty flavor score of the cheese made from goat milk.

Volatile Branched-Chain Fatty Acids (BCFA)

Ruminant fatty acids include straight-chain saturated, unsaturated, heteroatomic, and branched-chain fatty acids (Jensen, 2002; Schroeder et al., 2013). Branched-chain fatty acids represent about 1.4%-2.4% of the milk fatty acids. Volatile branched-chain fatty acids (BCFA), exhibiting branching at 4-position, mainly 4-methyloctanoic acid (4-Me-8:0), 4-ethyloctanoic acid (4-Et-8:0), and 4-methylnonanoic acid (4-Me-9:0) (Figure 1) have goaty/muttony/sheepy aroma and may be key flavor compounds of goat milk (Brennand, Ha, & Lindsay, 1989; Salles et al., 2002). Qualitative and quantitative profiles for most BCFA were similar in cow and goat milk except free fatty acids 4-Me-8:0 and especially 4-Et-8:0 are rare or not present in cow milk (Salles et al., 2002). These minor saturated compounds are derived from the lipolysis of triglycerides and are in the low amount (Cousin, 1982). 4-Me-8:0 in goat milk varies from 210-330 while 4-Et-8:0 from 140-170 μg/g fat. 4-Me-9:0 in goat milk is below 1 μg/g fat (Kaffarnik, Preuß, & Vetter, 2014). The flavor threshold levels of the free fatty acids at pH 2 are 4-Me-9:0 (0.65 ppm), 4-Me-8:0 (0.02 ppm) and 4-Et-8:0 (0.006 ppm) (Kaffarnik, Preuß, et al., 2014). 4-Me-8:0 exhibits the mutton-like aroma at concentrations below 100 ppb and blended easily with goaty aroma of 4-Et-8:0 to produce the goatiness (Attaie & Richter,

1996; Ha & Lindsay, 1990). Maturation and storage of goat cheese and yogurt may intensify the goaty flavor caused by an increase of the concentration of free volatile 4-alkyl-branched fatty acids (Kaffarnik, Preuß, et al., 2014). The content of BCFA is related to genetics, animal diet, lactation period, season and industrial processes and so is the goaty flavor (Teng, Reis, Ma, & Li, 2018).

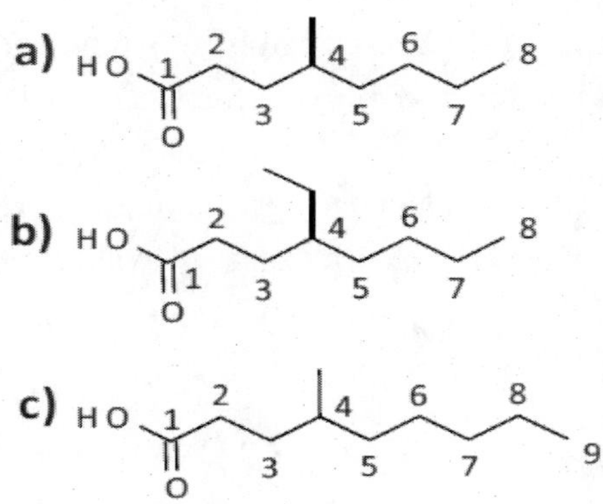

Figure 1. Chemical structures of (a) 4-methyloctanoic acid (4-Me-8:0), (b) 4-ethyloctanoic acid (4-Et-8:0), and (c) 4-methylnonanoic acid (4-Me-9:0).

MEASUREMENT OF GOATY FLAVOR

Sensory Evaluation

Analysis of characteristic food flavors has commonly been carried out by human assessment. For sensory evaluation of goaty flavor, taste and aroma are assessed by trained sensory panels. Sensory panel are asked to evaluate the attributes of the goat milk or dairy products samples usually with a scale from 0 (none) to 10 (very strong) (Yang, Ding, Ma, & Jia, 2015). Before evaluation, samples in a bottle with a lip are kept at a certain temperature for the development of flavor in the headspace. During evaluation, panelists are instructed to open the lid and take 3 short sniffs and taste before giving the final score.

Electronic Nose

Electronic nose is a potentially useful tool to evaluate goaty flavor intensity in goat milk samples. An electronic nose based on metal oxide sensors was used to measure goaty flavor in goat milk samples (Yang et al., 2015). The determination was carried out as following: goat milk samples were equilibrated at a certain time allowing for the development of headspace and then the flavor was determined by one needle used to absorb the air accumulated inside a sample vial. Sensor response expressed as the ratio of conductance of the sensor before and after exposure to the gas samples was obtained and analyzed by a computer. The electronic nose system evaluation correlated well with sensory evaluation with a coefficient of determination (R^2) of the multivariate partial least square regression (90%).

Gas Chromatography with Mass Spectrometry (GC-MS)

The free fatty acid in goat milk and its relative dairy products can be determined by a Gas Chromatography with Mass Spectrometry (GC-MS) (Kaffarnik, Kayademir, Heid, & Vetter, 2014). Briefly, the free fatty acids in the samples were generally extracted using a head-space solid-phase extraction method at certain temperature and for certain time. Then, the fiber was desorbed in the GC injection port and the fatty acids are measured using a GC-MS. For only determination of 4-Me-8:0, 4-Et-8:0 and 4-Me-9:0, which are key-flavor compounds but in low amounts, a method was developed using a Gas Chromatography-Mass Spectrometry Selected Ion Monitoring (GC-MS-SIM) (Kaffarnik, Preuß, et al., 2014). Goat milk or its products were lyophilized and extracted for milk fat using accelerated solvent extraction (ASE) and then the milk fat extract was converted to methyl esters. Identification of these volatile branched-chain fatty acids was obtained by comparing the retention time in the external standard and samples.

FACTORS INFLUENCING GOATY FLAVOR

Goat milk FFA composition and lipolytic process play an important role in the development of goat flavor (release of FFA). The lipoprotein lipase (LPL) activity is a very important reason for increasing flavor intensity. Goat milk lipolysis and LPL activity vary considerably and in parallel across goat breeds or genotypes, and are low during early and late lactation, as well as when animals are underfed or receive a diet supplemented with protected or unprotected vegetable oils (Chilliard et al., 2003). Thus, the goaty flavor of goat milk and its products differs depending on many factors such as genetics, lactation period and diet (Skjevdal, 1979).

Ricotta cheese made from whey of Sirina and Maltese goat milk were more goaty than those made from Girgentana and local breeds in Italy (Pizzillo, Claps, Cifuni, Fedele, & Rubino, 2005). Uysal-Pala, Karagul-Yuceer, Pala and Savas (2010) compared the sensory properties of drinkable yogurts made from milk of Turkish Saanen, Maltese and Turkish Hair goats and found that samples made from Turkish Saanen breed had more goaty flavors than those made from the other two breeds.

Flavor intensity of goat milk changes during lactation as FFA content varies. Lowest flavor intensities were found at the beginning and toward the end of the lactation cycle (Skjevdal, 1979). Maximum FFA content was found during mid-lactation (Agnihotri & Prasad, 1993). However, Uysal-Pala et al. (2010) found that drinkable yogurts produced from milk at the end lactation period had higher intensity of goaty flavor than those prepared from goat milk during the beginning and middle lactations.

Feeding regime to goats has an important effect on the medium-chain fatty acid composition (C_{6-14}) (Alvarez, Zarco, Galindo, Blache, & Martin, 2007). Feeding a concentrate rich in rapeseed oil improves fatty acid composition and flavor in Norwegian goat milk (Teng et al., 2018). Supplementation with 3% and 5% of castor oil led to a slightly stronger goatish odor of goat milk (Pereira et al., 2010).

Increase in free fatty acid can be seen during the processing of milk upon several treatments. It has been suggested that the final level of FFA in pasteurized milk at around 60% to 70% is due to lipolysis during milking and milk transfer to the bulk tank (Anderson, 2010). During a study of milk lipoprotein lipase in the mammary gland of 5 goats, three responded to hourly milking with a fivefold increase in LPL activity (Azzara & Dimick, 1989). Ohmic pasteurization at 72°C for 15 s did not promote an extended modification of free fatty acid content in goat milk when compared with that of conventional pasteurization (Pereira, Martins, & Vicente, 2008). For goat milk cheese, ripening time significantly affected the concentrations of FFA; concentrations increased during the initial 12 week of aging and remained relatively constant for the rest of the 24week ripening period. Therefore, the intensity and sharpness of flavor that originated from or was enhanced by the presence of these FFA did not change greatly after 12 weeks of aging (Attaie & Richter, 1996). Goat milk, goat yoghurt and goat cheese have different concentrations of 4-Me-8:0 and 4-Et-8:0 and showed different flavor (Kaffarnik, Kayademir, et al., 2014).

METHODS FOR MANAGING GOATY FLAVOR

In some regions, especially in China, the unique flavor of goat milk and its products may impact its acceptance to consumers and thus limit the market opportunities for goat milk. In recent years, various methods (Figure 2) have been developed to minimize the goaty flavor and improve the sensory properties of goat milk and its dairy product.

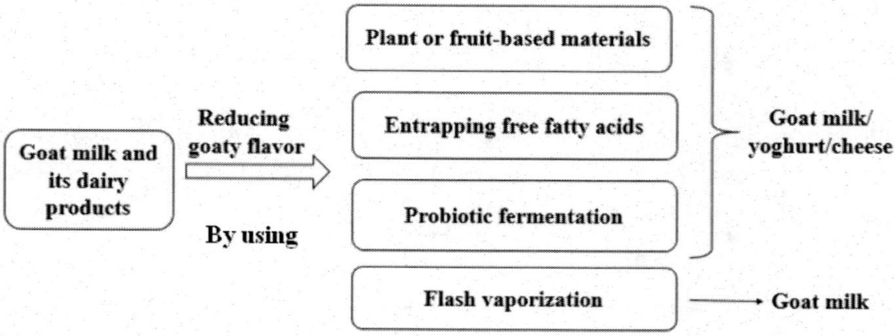

Figure 2. Common methods for managing goaty flavor of goat milk and its products.

Plant or Fruit-Based Materials

Incorporation of plant or fruit-based materials into the goat's milk yogurt may help to mask the characteristic unpleasant flavor and potentially increase acceptability among consumers. Addition of jujube pulp weakened the goaty flavor greatly, improving the sensory acceptance, and increased the antioxidant activities of goat milk yoghurt (Feng et al., 2019). Mangosteen (Garcinia mangostana Lin.) is one of the most favorite fruits in Indonesia. Addition of mangosteen rind extract (1%-4%, v/v) to goat milk yogurt drink improved the odor score from 2.12 to 3.60 and preference from 2.82 to 3.20. Different levels of mangosteen rind extract affected the formation of flavor compounds (Wibawanti, Rinawidiastuti, Arifin, & Zulfanita, 2018). Flavor of fermented goat milk was significantly increased when 2% (w/w) grape pomace extract was added (Santos et al. 2017). Incorporation of stingless bee honey (5-10%, v/v) positively affected the flavor and acceptability of goat milk yogurt (Machado et al, 2017). Mixed fruit flavor was added to the concentrated goat milk before spray drying to reduce the goaty flavor in the final powder (Reddy et al., 2014). A mixed fruit juice (apple juice (45%), orange juice (44%), banana puree (5%), pineapple juice (4%), mango puree (1.3%) and passionfruit juice (0.6%)) was added to plain goat milk yoghurt to make stirred yoghurts at the level of 5, 10 or 15% (w/w).

The authors reported that there were no complaints regarding the characteristic unpleasant goaty taste when the mixed fruit juice was added up to 10% and 15% and the reasons for the improved flavor was the incorporation of natural sugar through addition of fruit juice and combined contribution from flavor compounds in fruit juice and improved higher viability of *L. acidophilus* LA-5, which may also produce flavor compounds (Senaka Ranadheera et al., 2012).

Mixed dairy products (containing a percentage of cows' milk) may represent a means to increase the consumption of goat milk. Using a mixture of equal amount of goat and cow milks to produce drinkable yogurt improves the flavor by decreasing goaty flavor (Uysal-Pala et al., 2010).

Entrapping Agents

Cyclodextrins (CD) are macrocyclic compounds consisting of a variable number of glucose-molecule rings linked through alpha (1→4) bonds (Hanna & Kollengode, 2010). CD can be divided into α-, β-, and γ- types according to the number of glucose-molecule rings which is 6, 7, and 8, respectively. The cavity of β-cyclodextrin (β-CD, Figure 3) is hydrophobic and thus can bind hydrophobic molecules with suitable size. Studies shows that β-CD has the ability of trapping short-medium chain free fatty acids and therefore has been used to reduce goat milk flavor (Sadooghysaraby, 2011). β-CD forms inclusion complexes with free volatile octanoic and decanoic acids and changs them to be nonvolatile, which then alters the perception of goat milk flavor (Meier, Drunkler, Luiz, Fett, & Szpoganicz, 2001). In vitro studies of the inclusion of formed complexes between β-CD and short-chain fatty acids, caprylic (C_8) and capric acid (C_{10}), were carried out by Differential Scanning Calorimetry (DSC) and Proton Magnetic Resonance (1H NMR).

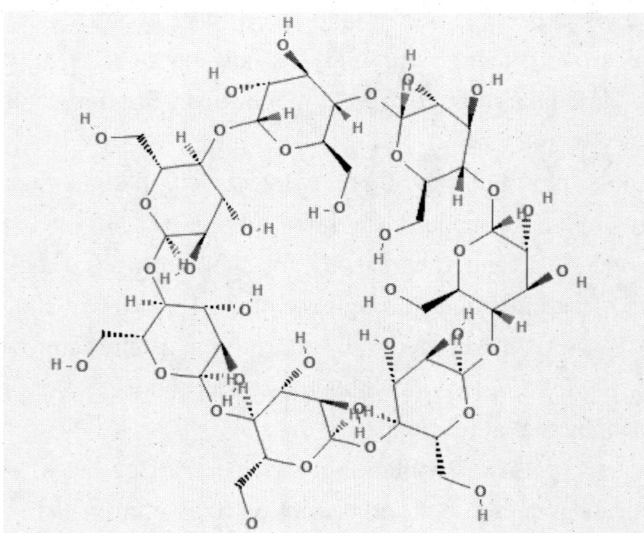

Source: PubChem, URL: https://pubchem.ncbi.nlm.nih.gov.

Figure 3. 2D Structure of β-cyclodextrins (β-CD).

Complex formation was confirmed by DSC, through the disappearance of the signs due to melting of C_{10} and C_8 after the encapsulation process. The 1H NMR technique shows that C10 is included in the β-CD cavity (Meier et al., 2001).

Affinity of straight and branched chain fatty acids to β-CD was also studied (Young, Gupta, & Sadooghy-Saraby, 2012) and the authors reported that β-CD bind more 4-methyl branch than straight chain isomers of octanoic acid in lipase-treated goat milk. β-CD is also effective in reducing goaty flavor of goat yoghurt. However, β-CD must be added before heating and fermentation.

Our previous study showed that polymerized whey protein can be used to reduce the goaty flavor of goat milk yogurt (Wang, Wang, Gao, Xu, & Guo, 2018). Polymerized whey protein (PWP) is soluble whey protein aggregates formed by heating whey protein at certain concentration under controlled conditions (Vardhanabhuti, Foegeding, Mcguffey, Daubert, & Swaisgood, 2001). PWP (0.7%, w/v) addition can decrease contents of fatty acids (C6, C8, C10) in goat milk yoghurt approximately by 22, 71, 54%, respectively (Wang, Wang, Gao, Xu, & Guo, 2018).

We assume that there may be two reasons for this effect. β-Lactoglobulin (β-LG, Figure 4), the main whey protein, is a member of lipocalin family and had great affinity towards fatty acid via hydrophobic interaction. It is composed of eight antiparallel β-strands arranged in a barrel (A–H) and one protective α-helix. There are two hydrophobic pockets in the molecules which in the cavity of the barrel and the other locate between the C-terminal of α-helix and barrel (Perez, Andermatten, Rubiolo, & Santiago, 2014).

Study showed that β-lactoglobulin can bind fatty acid ranging from 8 (short fatty acid) to 20 carbon atoms (long fatty acids) in the cavity of barrel (Stefania, Rita, & Bruno, 2015). A protein gel consists of a spatial network. After addition of polymerized whey protein into the goat milk, acid-induced cold-set gelation of PWP can be formed when pH is lowered to the isoelectric point of PWP during goat milk fermentation, where the electrostatic repulsion trends to decrease and consequently trap SM-FFA in the network (Wang, Bao, Hendricks, & Guo, 2012).

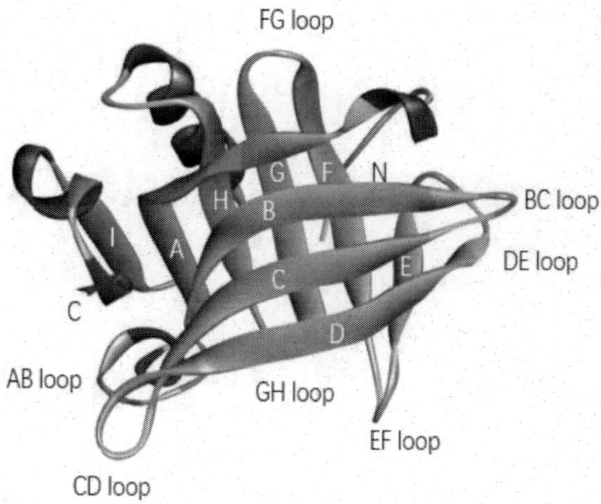

Figure 4. Monomeric unit of bovine β-lactoglobulin. The structure (3uew) was obtained from the Research Collaboratory for Structural Bioinformatics (RCSB) Protein Data Bank (http://www.rcsb.org).

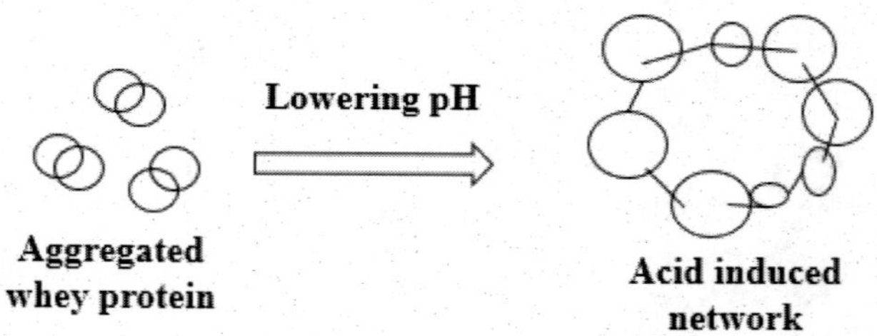

Figure 5. Network of acid-induced polymerized whey protein gel. Modified from Alting (2003).

The network (Figure 5) can entrap water and small molecules inside. Thus, the free fatty acid may be entrapped in the network and become nonvolatile.

Lactic Acid Fermentation

Using regular yogurt cultures to make yogurts significantly affected the intensity of goaty flavor. Goat milk yoghurt had more acceptable and sensory appealing to consumers due to the flavor compounds produced by fermentation (Uysal-Pala et al., 2010). Probiotics are living microorganisms which are beneficial to human health when administered in adequate amounts (FAO/WHO, 2001). Incorporation of probiotic confers more sensory advantage to the final products (Vinderola, Bailo, & Reinhemier, 2000).

The main mechanism is due to the flavor compounds formation by degradation of protein, lactose and fat into such as diacetyl, acetoin, acetaldehyde, or acetic acid, methyl ketones, secondary alcohols, esters and lactones (Bertuzzi et al., 2017) which can cover the goaty flavor. A set-type fermented milk was developed by incorporation of 3% whey protein concentrate which can mask the characteristic taste of goat's milk. In addition to the role whey protein concentrate played, it should be noticed that probiotic *Bifidobacterium* BB-12 was increased from 3.6 to 16.6×10^6 (MartíN-Diana, Janer, Peláez, & Requena, 2003). *Lactobacillus acidophilus* LA-5 may also produce flavor compounds, such as acetaldehyde, which are recognized as important flavor components in yogurt. Thus, high viability of *L. acidophilus* LA-5 in goat milk yoghurt can reduce the goaty flavor (Senaka Ranadheera et al., 2012). Flavor of goat fermented milk containing *L. rhamnosus* HN001 (Santos et al. 2017) or *L. Acidophilus* La-5 (Machado et al., 2017) were significantly improved.

Other Methods

Besides the commonly used methods mentioned above, other methods such as transgenic goats and flash vaporization have also been developed. Baldassarre et al. (2008) introduced human butyrylcholinesterase in the mammary gland in the transgenic goat and found that milk fatty acid composition for transgenic goats, as determined by Gas Chromatography, was significantly fewer short chain fatty acids. Flash vaporization is a commonly used dairy industrial technique. It removes the volatile free fatty acid and other off-flavor compounds by vacuum evaporation and then cooling using the principle of decreasing boiling point by decreasing pressure.

Conclusion

Goaty flavor of goat milk and its products plays a significant role in new products development and marketing. Short and medium chain fatty acid including those volatile branched fatty acids may be responsible for this unique flavor. Physical, chemical, microbiological and genetic methods may be used to improve the flavor property of goat milk and its products.

References

Alonso, L., Fontecha, J., Lozada, L., Fraga, M. & Juárez, M. 1999. Fatty acid composition of caprine milk: major, branched-chain, and trans fatty acids. *Journal of Dairy Science,* 82, 878-884.

Alting, A. C. 2003. *Cold gelation of globular proteins.* Wageningen University, 1591, 186-188.

Alvarez, L., Zarco, L., Galindo, F., Blache, D. & Martin, G. B. 2007. Social rank and response to the "male effect" in the Australian Cashmere goat. *Animal Reproduction Science,* 102, 258-266.

Anderson, M. 2010. Milk lipase and off-flavour development. International *Journal of Dairy Technology,* 36, 3-7.

Attaie, R. & Richter, R. L. J. J. O. D. S. 1996. *Formation of volatile free fatty acids during ripening of Cheddar-like hard goat cheese.* 79, 717-724.

Azzara, C. D. & Dimick, P. S. 1989. Paracellular leakage of lipoprotein lipase across the mammary epithelium of the goat 1. *Journal of Dairy Science, 72*, 1159-1168.

Baldassarre, H., Hockley, D. K., Olaniyan, B., Brochu, E., Zhao, X., Mustafa, A. & Bordignon, V. 2008. Milk composition studies in transgenic goats expressing recombinant human butyrylcholinesterase in the mammary gland. *Transgenic Research,* 17, 863-872.

Bertuzzi, A. S., Kilcawley, K. N., Sheehan, J. J., O'sullivan, M. G., Kennedy, D., Mcsweeney, P. L. H. & Rea, M. C. 2017. Use of smear bacteria and yeasts to modify flavour and appearance of Cheddar cheese. *International Dairy Journal,* S095869461730078X.

Brennand, C. P., Ha, J. K. & Lindsay, R. C. 1989. Aroma properties and thresholds of some branched-chain and other minor volatile fatty acids occurring in milkfat and meat lipids 1. *Journal of Sensory Studies,* 4, 105-120.

Chilliard, Y., Ferlay, A., Rouel, J. & Lamberet, G. 2003. A review of nutritional and physiological factors affecting goat milk lipid synthesis and lipolysis1. *Journal of Dairy Science,* 86, 1751-1770.

Costa, M., Balthazar, C., Franco, R., Mársico, E., Cruz, A. & Junior, C. C. 2014. Changes on expected taste perception of probiotic and conventional yogurts made from goat milk after rapidly repeated exposure. *Journal of Dairy Science,* 97, 2610-2618.

Cousin, M. 1982. Presence and activity of psychrotrophic microorganisms in milk and dairy products: a review. *Journal of Food Protection,* 45, 172-207.

FAO/WHO. Food and Agriculture Organization of the United Nations; World Health Organization (2001). *Evaluation of health and nutritional properties of probiotics in food including powder milk with live lactic acid bacteria* (Córdoba, 34 pp.).

Feng, C., Wang, B., Zhao, A., Wei, L., Shao, Y., Wang, Y., Cao, B. & Zhang, F. 2019. Quality characteristics and antioxidant activities of goat milk yogurt with added jujube pulp. *Food Chemistry,* 277, 238-245.

Ha, J. K, Lindsay, R. C. 1990. Distribution of volatile branched-chain fatty acids in perinephric fats of various red meat species. *Lebensm Wiss Technol* 23: 433–40.

Hanna, M. A. & Kollengode, A. N. 2010. Cyclodextrin complexed flavors retention in extruded starches. Journal of Food Science, 62, 1057-1060.

Jensen, R. G. 2002. The composition of bovine milk lipids: January 1995 to December 2000. *Journal of Dairy Science,* 85, 295-350.

Kaffarnik, S., Kayademir, Y., Heid, C. & Vetter, W. 2014a. Concentrations of volatile 4-alkyl-branched fatty acids in sheep and goat milk and dairy products. *Journal of Food Science,* 79, C2209-C2214.

Kaffarnik, S., Preuß, S. & Vetter, W. 2014b. Direct determination of flavor relevant and further branched-chain fatty acids from sheep subcutaneous adipose tissue by gas chromatography with mass spectrometry. *Journal of Chromatography A,* 1350, 92-101.

Machado, T. A. D. G., De Oliveira, M. E. G., Campos, M. I. F., De Assis, P. O. A., De Souza, E. L., & Madruga, M. S., et al. (2017). Impact of honey on quality characteristics of goat yogurt containing probiotic, lactobacillus acidophilus. *LWT - Food Science and Technology,* 80, 221-229.

Martín-diana, A. B., Janer, C., Peláez, C. & Requena, T. 2003. Development of a fermented goat's milk containing probiotic bacteria. *International Dairy Journal,* 13, 827-833.

Meier, M. M., Drunkler, D. A., Luiz, M. T. B., Fett, R. & Szpoganicz, B. 2001. The influence of β-cyclodextrin on goaty flavour-Characterization of synthetic inclusion complexes with capric acid and caprylicacid. *British Food Journal,* 103, 281-290.

Mowlem, A. 2005. Marketing goat dairy produce in the UK. *Small Ruminant Research,* 60, 207-213.

Park, Y. & Drake, M. 2005. Effect of 3 months frozen-storage on organic acid contents and sensory properties, and their correlations in soft goat milk cheese. *Small Ruminant Research,* 58, 291-298.

Park, Y., Juárez, M., Ramos, M. & Haenlein, G. 2007. Physico-chemical characteristics of goat and sheep milk. *Small ruminant research,* 68, 88-113.

Pereira, R. A. G., Oliveira, C. J. B., Medeiros, A. N., Costa, R. G., Bomfim, M. A. D. & Queiroga, R. C. R. E. 2010. Physicochemical and sensory characteristics of milk from goats supplemented with castor or licuri oil. *Journal of Dairy Science,* 93, 456-462.

Pereira, R. N., Martins, R. C. & Vicente, A. A. 2008. Goat Milk Free Fatty Acid Characterization During Conventional and Ohmic Heating Pasteurization. *Journal of Dairy Science,* 91, 2925-2937.

Perez, A. A., Andermatten, R. B., Rubiolo, A. C. & Santiago, L. G. 2014. β-Lactoglobulin heat-induced aggregates as carriers of polyunsaturated fatty acids. *Food Chemistry,* 158, 66-72.

Pizzillo, M., Claps, S., Cifuni, G. F., Fedele, V. & Rubino, R. 2005. Effect of goat breed on the sensory, chemical and nutritional characteristics of ricotta cheese. *Livestock Production Science,* 94, 33-40.

Posati, L. P., Orr, M. L., 1976. Composition of Foods, Dairy and Egg Products, *Agriculture Handbook No. 8-1.* USDA-ARS, Consumer and Food Economics Institute Publishers, Washington, DC, pp. 77–109.

Ranadheera, C. S., Evans, C. A., Adams, M. C. & Baines, S. K. 2012. Probiotic viability and physico-chemical and sensory properties of plain and stirred fruit yogurts made from goat's milk. *Food Chemistry,* 135, 1411-1418.

Ranadheera, C. S., Naumovski, N. & Ajlouni, S. 2018. Non-bovine milk products as emerging probiotic carriers: Recent developments and innovations. *Current Opinion in Food Science.*

Reddy, R. S., Ramachandra, C. T., Hiregoudar, S., Nidoni, U., Ram, J. & Kammar, M. 2014. Influence of processing conditions on functional and reconstitution properties of milk powder made from Osmanabadi goat milk by spray drying. *Small Ruminant Research,* 119, 130-137.

Ribeiro, A. & Ribeiro, S. 2010. Specialty products made from goat milk. *Small Ruminant Research,* 89, 225-233.

Sadooghysaraby, S. 2011. Assessment of the entrapment of free fatty acids in goat milk by β-cyclodextrin and reduction of goaty flavour. *Auckland University of Technology.*

Salles, C., Sommerer, N., Septier, C., Issanchou, S., Chabanet, C., Garem, A. & Quéré, J. L. L. 2002. Goat cheese flavor: Sensory evaluation of branched-chain fatty acids and small peptides. *Journal of Food Science,* 67, 835-841.

Santos, K. M. O., Bomfim, M. A. D., Vieira, A. D. S., Benevides, S. D., Saad, S. M. I., Buriti, F. C. A., & Egito, A. S. (2012). Probiotic caprine Coalho cheese naturally enriched in conjugated linoleic acid as a vehicle for Lactobacillus acidophilus and beneficial fatty acids. *International Dairy Journal,* 24, 107-112.

Schroeder, Markus, Vetter & Walter 2013. Detection of 430 Fatty Acid Methyl Esters from a Transesterified Butter; Sample. *Journal of the American Oil Chemists Society,* 90, 771-790.

Skjevdal, T. 1979. Flavour of goat's milk: A review of studies on the sources of its variations. *Livestock Production Science,* 6, 397-405.

Stefania, E., Rita, G. & Bruno, R. 2015. Molecular simulations of β-lactoglobulin complexed with fatty acids reveal the structural basis of ligand affinity to internal and possible external binding sites. *Proteins-structure Function & Bioinformatics,* 82, 2609-2619.

Teng, F., Reis, M. G., Ma, Y. & Li, D. 2018. Effects of Season and Industrial Processes on Volatile 4-Alkyl-Branched Chain Fatty Acids in Sheep Milk. *Food Chemistry,* 260, 327.

Uysal-pala, C., Karagul-yuceer, Y., Pala, A. & Savas, T. 2010. Sensory properties of drinkable yogurt made from milk of different goat breeds. *Journal of Sensory Studies,* 21, 520-533.

Vardhanabhuti, B., Foegeding, E. A., Mcguffey, M. K., Daubert, C. R. & Swaisgood, H. E. 2001. Gelation properties of dispersions containing polymerized and native whey protein isolate. *Food Hydrocolloids,* 15, 165-175.

Vinderola, C. G., Bailo, N. & Reinhemier, J. A. 2000. Survival of probiotic microflora in Argentinian yoghurts during refrigerated storage. *Food Research International*, 33, 97-102.

Wang C., W., Wang, C. N., Gao F., Xu Y. Y. & Ruo M. G. Effects of polymerized whey protein on goaty flavor and texture properties of fermented goat milk in comparison with β-cyclodextrin. *Journal of Dairy Research.*

Wang, W., Bao, Y., Hendricks, G. M. & Guo, M. 2012. Consistency, microstructure and probiotic survivability of goats' milk yoghurt using polymerized whey protein as a co-thickening agent. *International Dairy Journal,* 24, 113-119.

Wibawanti, J. M. W., Rinawidiastuti, Arifin, H. D. & Zulfanita 2018. Improving characteristics of goat milk yogurt drink fortified by mangosteen rind (Garcinia mangostana Lin.) extract. In: ALBAARRI, A. N. (ed.) *International Symposium on Food and Agro-Biodiversity.*

Wilson, A. & Baietto, M. 2009. Applications and advances in electronic-nose technologies. *Sensors,* 9, 5099-5148.

Yang, C., Ding, W., Ma, L. & Jia, R. 2015. Discrimination and characterization of different intensities of goaty flavor in goat milk by means of an electronic nose. *Journal of Dairy Science,* 98, 55-67.

Young, O. A., Gupta, R. B. & Sadooghy-Saraby, S. 2012. Effects of cyclodextrins on the flavor of goat milk and its yogurt. *Journal of Food Science,* 77, S122-S127.

In: Goat Milk Chemistry …
Editor: Mingruo Guo
ISBN: 978-1-53617-299-7

Chapter 6

GOAT MILK PRODUCTS AND THEIR MANUFACTURING TECHNOLOGY

Shilong Jiang[1], Man Su[1], Mu Wang[2], Hao Wang[2] and Mingruo Guo[2,3,*]
[1]Feihe Dairy Company, Harbin, Heilongjiang, China
[2]Key Laboratory of Dairy Science, Northeast Agricultural University, Ministry of Education, Harbin, Heilongjiang, China
[3]Department of Nutrition and Food Sciences, University of Vermont, Burlington, Vermont, US

ABSTRACT

Goats play a vital role in improving economic efficiency in tropical and arid regions of developing countries. Compared with cow milk, goat milk is less allergenic due to the lower αs-1 casein content, which makes it more popular all over the world. This chapter focuses on the main products of goat milk including liquid goat milk, goat milk powder, goat milk butter, fermented goat milk and goat milk cheese, as well as their processing technology and research development. By getting rid of the stale and bring

* Corresponding Author's Email: mguo@uvm.edu.

forth the fresh, with continuous emphasis on varieties and new techniques of goat milk products, the market and research for goat milk will grow faster in the future.

Keywords: goat milk product, technology, research, development

INTRODUCTION

Goats produce approximately 3.6% of the world's whole fresh milk supply annually (FAO, 2016). Goats play a vital role in improving economic efficiency in tropical and arid regions of developing countries. Among approximate 617 million goats in the world, 97.3% of them are in developing countries. of these, 65.9% are in Asia, 27.4% in Africa, 3.5% in Europe and 3.0% in the Americas. There are 191 million dairy goats in the world in 2012 and 47.7% are located in the 25 least developed countries (Escareño et al., 2012).

Goat milk is widely used to manufacture different types of products worldwide, including milk beverages, powdered products, butter, goat milk yogurt, soft cheeses, blue cheese and hard cheeses, etc. However, only cheese has been studied extensively among all goat milk products mentioned above (Loewenstein, Speck, Barnhart, & Frank, 1980). In the UK, approximate 15 to 20 million liters of goat milk are processed for consumption each year, of which 60% are for cheese, 20% for fluid milk, 10% for yogurt, and the rest for butter, cream and ice cream. Liquid milk products include whole, skimmed, semi-skimmed and UHT milk (Ribeiro & Ribeiro, 2010).

LIQUID GOAT MILK

A number of different liquid goat milk products are available on the market. High density of certain nutrients has made goat milk more and more popular. The process and packaging vary widely. There are a variety of fluid milks around the world, such as whole milk or skimmed milk, etc. Standards of goat milk in different countries are not the same, some countries focus on vitamin inclusion, and the others are on packaging size (from 454 grams to 3.6 liters) or packaging materials (plastic, carton or glass). Meanwhile, according to processing, packaging type and specific regulations, there is a wide range of shelf life in liquid goat milk (Ribeiro & Ribeiro, 2010). Safety is the leading reason for consumption, which is highly connected to heat treatment. Therefore, heat treatments (pasteurized or UHT) are necessary for liquid goat milk. Products should be delivered in a cold chain for milk with pasteurized treatment, while in ambient temperature for UHT treatment. Many countries prohibit or strictly restrict the delivery of raw milk. Likewise, delivering untreated milk may involve health hazards (Walstra, Wouters, & Geurts, 2006).

Another factor that influences consumption of liquid milk is the flavor. For example, cooked flavor is not a pleasant smell for western consumers. Low-temperature heat treatment is generally preferred in these countries. While in China, people prefer warm milk and care more about shelf life, therefore UHT milk is favored. When it comes to goat milk, the "goaty flavor" is unacceptable by most consumers. The ratio of short chain fatty acids (such as capric, caproic and caprylic acids) to all free fatty acids in goat milk is 20 to 25%. These short-chain fatty acids are responsible for "goaty" flavor (Young, Gupta, & Sadooghy-Saraby, 2012). While there are many applied approaches (breeding and production) to reduce goat milk flavor, so that the final product could be accepted by customers (Chilliard, Ferlay, Rouel, & Lamberet, 2003; Fedele, Rubino, Claps, Sepe, & Morone, 2005).

Pasteurized Goat Milk

Limited work has been done regarding fluid goat milk manufacture process. Such research is highly desirable, because there is likely to be variation in flavor quality of the raw milk supply, and goat milk may be susceptible to off-flavor and other quality issues due to many factors. The purpose of pasteurization is as follows (Claeys et al., 2013; Horner, Wallen, & Caporaso, 1980; Mchenry, 1934):

- Killing all pathogenic microorganisms that might be present in the raw milk;
- Reducing the number of bacteria that cause spoilage at refrigerated temperatures so that the packaged products can be stored for a week or more when refrigerated;
- Maintaining flavor, nutritional value and other properties as close as possible to fresh raw milk. The more intense the heat treatment, the more changes in these properties will occur for the processed milk are from raw milk.

Regarding the safety of pasteurized milk, most countries have legal requirements for the maximum number of microorganisms and E. coli and the lack of phosphatase activity. In order to meet these requirements, raw milk should not contain high level of heat-resistant bacteria, pasteurization should be sufficient, and re-contamination of pasteurized milk and microorganisms should be minimized.

Manufacturing Process

Liquid goat milk manufacturing process is similar to bovine milk, including milk collection, filtering, standardization, homogenization, cooling, packaging, storage and delivering. Figure 1 shows an example of the manufacture of pasteurized milk.

Heat treatment is needed to prevent fat and protein breakdown by heat-resistant enzymes of psychrotrophic bacteria. Usually, pasteurized milk is stored for too short a time to cause significant decomposition by these enzymes, unless raw milk contains a large number of psychrophilic bacteria. In addition, the significant inactivation (about 50%) of lipase was caused by heating at a relatively high temperature (e.g., 67.5°C for 20 s) and a slightly lower pasteurization temperature was allowed in homogenized milk production. Despite these obvious advantages, dairy factories usually only cool the milk (mainly to save on costs) at the risk of the growth of psychrotrophic bacteria.

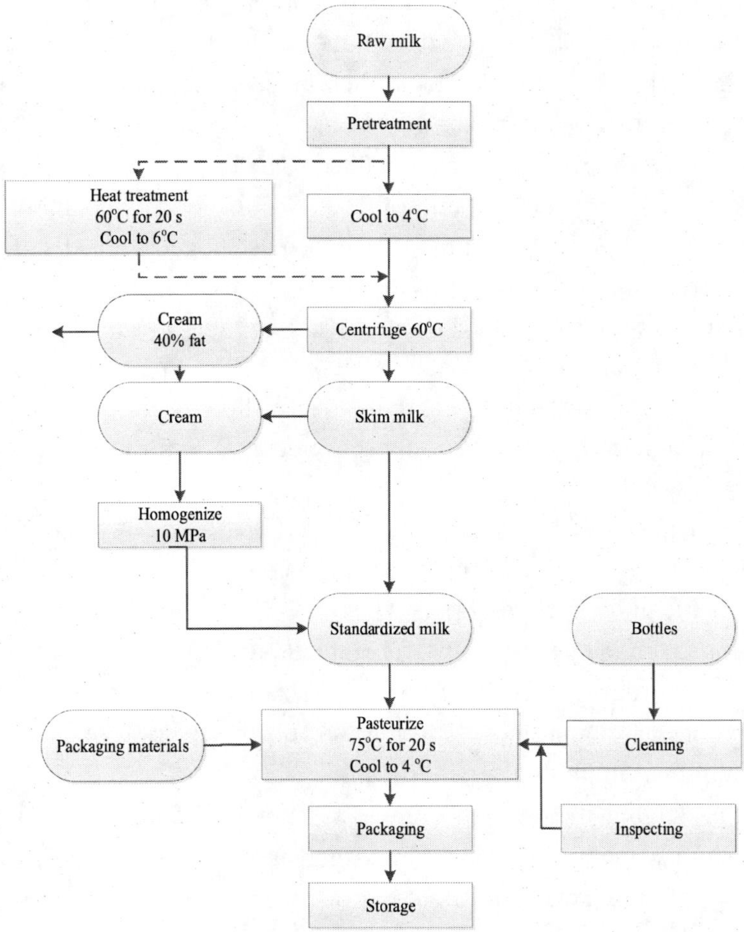

Figure 1. Flow chart of pasteurized liquid milk processing.

Cream separation is needed to standardize the milk. If homogenization is omitted, only a part of the milk is skimmed, and the amount of skimmed milk obtained should be sufficient to standardize the milk. However, cream separation is optional in goat milk, and goat milk cream is seldom on the market. Except for standardizing the goat milk, separation also serves to manufacture butter, ice cream or ingredient.

Homogenization serves to prevent the formation of cream layers during storage. Homogeneous milk has an increased tendency to foam, especially at low temperatures. Standardization can be accomplished by adding skimmed milk (or cream) to the milk or by continuous standardization. As a result, not all the milk is homogenized but only its cream fraction, to reduce the costs. Obviously, all milk should then be separated. Therefore, the fat content of the cream should be adjusted to 10-12% of skimmed milk, and the temperature should not be too low (≥ 55°C). Homogenization is usually done prior to pasteurization to minimize the risk of recontamination. Since milk lipase is still present, the milk should be pasteurized immediately after homogenization.

Pasteurization ensures safety and greatly extends the shelf life of the product. Heat treatment at 72°C for 15 seconds kills all pathogens that may be present and leads phosphatase activity inactivation. A more intensive heat treatment is usually applied because the homogenized milk is very sensitive to lipolysis.

Pasteurization does not inactivate bacterial lipases and proteases. These enzymes will not cause problems because pasteurized milk has a relatively short shelf life under refrigeration.

During manufacture of low temperature pasteurized milk, plate heat exchangers are universally used for pasteurization. The time-temperature combination is a compromise between the full inactivation of milk lipase and the ability to inhibit bacterial growth.

The packaging of low temperature pasteurized bottle milk is usually carried out in a single service container, such as a carton or plastic. In terms of product safety, special care should be taken to ensure hygiene during the packaging process, especially since the shelf life of the product is again affected; aseptic packaging is ideal. The packaging material is sterilized by a combination of hydrogen peroxide and heat or ultraviolet light in an aseptic filling machine.

Shelf Life

Shelf life is the time during which the pasteurized product can be kept under conditions (e.g., at a given temperature) without apparent undesirable changes. The shelf life of pasteurized milk varies greatly in different countries and regions. Pasteurized milk has a shelf life for only a couple of days in some countries to over 20 days in others. The reasons for the variation are raw milk quality, processing conditions, and last but not least the temperature during storage.

Deterioration of pasteurized milk is mainly caused by growth of microorganisms. It is determined by:

- the number of bacteria that survive pasteurization
- pasteurization conditions
- extent of recontamination
- storage temperature
- activity of substances inhibiting bacterial growth

The storage temperature of milk is important because the microbial production time is highly temperature dependent. It does not make sense to reduce the temperature below 4 to 5°C, because during the distribution and storage of the distribution network, higher temperatures usually occur, such as 7°C (Griffiths & Phillips, 2010).

Strict aseptic filling conditions and frequent and thorough inspections are required during processing to limit recontamination and meet the requirements of the final sale day. For this purpose, the sample can be held at various temperatures and tested at intervals. The disadvantage is that the user has received milk before the known shelf life test results. Therefore, tests have been developed that allow for relatively rapid recontamination detection by Gram-negative, non-spore forming bacteria.

In addition to the hygienic state of the package, packaging may also be a factor in shelf life depending on its ability to protect milk from oxygen and light. The changes in nutritional (i.e., destruction of vitamins) and sensory properties may occur during storage (Petrus, Loiola, & Oliveira, 2010;

Simon & Hansen, 2001). Pasteurized milk has a high initial oxygen content, and most packages have a considerable oxygen content in the headspace. When the milk is contained in a transparent bottle, the presence of oxygen becomes important under the action of light. The resulting off-flavor is caused by the oxidation of sulfur-containing amino acids and the oxidation of lipids. Vitamin B2 is responsible for these photoinduced oxidation reactions, acting as a photocatalyst for oxidation.

Ultra-High Temperature Treatment (UHT) Goat Milk

Ultra-high temperature treatment (UHT) of milk is a heating process for production of a commercially sterile product, that is, a product in which bacteria will not grow for extended period of time when packed aseptically and stored at room temperature. It involves heating at a temperature higher than 135°C with a holding time of $<$1-10 s. The undesirable side effects of conventional sterilization such as browning, bactericidal taste and vitamin loss are reduced by UHT treatment. Certain enzymatic and physico-chemical changes may still occur after UHT sterilization.

A cooperative dairy company in Perugia has produced goat UHT milk in Italy. Milk is sold in a 500-mL carton containing 60 kcal, 2.8 g protein, 4.6 g carbohydrate, 3.0 g fat and 119 mg calcium per 100 mL. Commercial UHT goat milk is also available in the US and other countries.

According to Australian standards, alkaline phosphatase tests for monitoring the efficiency of milk pasteurization may not be suitable for certain products and are not applicable to goat milk or goat milk products, nor to UHT products (Pandya & Ghodke, 2007).

Heat Stability of Goat Milk

Goat milk may be sensitive to extensive heat and therefore cannot withstand ultra-high temperature treatment (Fox & Hoynes, 1976; Zadow, Hardham, Kocak, & Mayes, 1983). However, individual milk heat treatments vary widely. Heat coagulation time at 140°C is 0.5-23.4 minutes. The thermal coagulation temperatures of different goat milks vary widely. The range is from 118°C to above 140°C (Fox & Hoynes, 1976). The so-called thermal solidification temperature means that the milk does not exceed 1 minute at this temperature, otherwise the temperature at which solidification occurs (Remeuf, 1992).

This thermal instability of goat milk is due to its pH threshold of milk. The threshold pH for goat milk is approximately at 6.9 (Zadow et al., 1983). Goat milk with a starting pH of 6.7 shows lower thermal stability than bovine milk. The difference in micelle characteristics or salt balance of goat milk may also cause its lower thermal stability (Remeuf, 1992; Zadow et al., 1983). This explanation is based on the fact that at the time of heat treatment, a higher ion concentration of calcium and a lower micelle solvent cause instability of the micelles. There are several ways to improve the thermal stability of goat milk, including adjusting the pH, increasing the calcium chelating agent and preheating the milk (Remeuf, 1992; Zadow et al., 1983).

POWDERED GOAT MILK PRODUCTS

The main purpose of manufacture of milk powder is to convert liquid perishable raw material to a product that can be stored without substantial loss of quality, preferably for some years. The powder should be free-flowing, i.e., flow readily from an opening, and not stick to the walls of vessels and production equipment. The latter requirement is of special importance for powder used in, e.g., coffee machines. In addition, after adding water, the powder should be reconstituted to a homogeneous mixture, similar in composition to original product.

The composition of the powder also varies due to the composition of the raw materials. This provides a possibility for adulteration; for example, buttermilk powder or whey powder can be added to (defatted) milk powder. The presence of most exotic powders can be detected under the microscope, but it is usually not easy to determine the mixing of a small amount of another liquid before drying. Because whey is cheaper than skim milk, this "adulteration" sometimes occurs.

Goat Milk Powder

There is very little literature on the use of goat milk to make powders, probably because a large amount of goat milk from a large number of small production farms is not available, and goat milk is often mixed with bovine milk. Three types of drying methods in goat milk powder making are freeze drying (Shurpalekar, Chandrasekhara, Swaminathan, Sreenivasan, & Subrahmanyan, 1964), roller drying, and spray drying. As we all know, freeze dried skimmed milk powder plays an important role in goat milk standardization (Loewenstein, Speck, & Barnhart, 1979). There are about 3.1% moisture, 96.9% total solids, and 1.2% fat contained in skimmed goat milk powder. And for roller and spray-dried powders were 3.5, 96.5, 0.98%, and 2.84, 97.2 and 0.99%, respectively. The information about manufacturing skim goat milk powder is very limited.

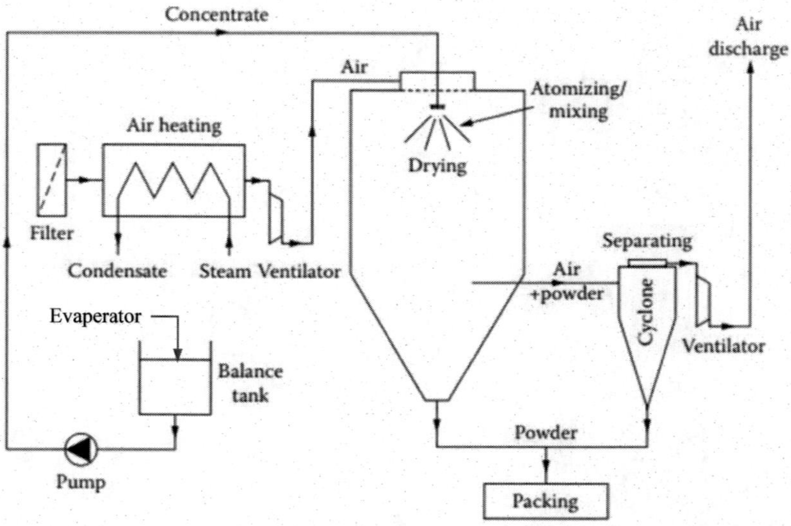

Figure 2. Simplified diagram of spray drying process. Adapted from Walstra et al. (2006).

Spray drying is the mostly used method in milk powder making. In order to protect the milk solids away from damaging by heating process, low temperature should be chosen during milk spray drying. Spaying the concentrate as very fine droplets by a nozzle or a rotary atomizer into a hot air stream can reduce the demand for temperature. Simplified diagram of spray-drying process has been shown in Figure 2.

Manufacturing Process

Goat milk powder production process is similar to bovine milk powder. Figure 3 gives an overview of manufacture of whole milk powder.

Preheating is needed for a set time/temperature combination to obtain a desired characteristic powder. Preheating process, such in low, medium or high temperature, should be included in the extent of serum protein denaturation temperature. The extent of serum protein denaturation has a significant influence on the functionality of the milk powder in its end use.

During nozzle atomization, the fat globules are effectively disrupted in the nozzle. However, homogenization is usually applied to reduce the free fat content. Homogenization of high concentration milk significantly increases its viscosity. This increase results in the formation of coarser droplets during atomization. Therefore, dry matter content of powder maybe high in no homogenized milk.

The milk is concentrated by evaporation to remove water. In order to get a high content of total solids before drying, thermal evaporation maybe a less expensive process by water removing.

In order to prevent the growth of microorganisms, cold storage of concentrated milk before spray drying is necessary. However, cold concentrated milk has a higher value of viscosity which makes it difficult to be atomized. A heating process must be applied before spray drying.

Gas flushing by using N_2 or mixture of N_2 and CO_2 instead of air could decrease the autoxidation and it can help to improve the stability of the powder. It can be done once or twice. It is applied in infant formula containing high amounts of unsaturated fatty acids but not often in whole milk powder. The initial oxygen content is reduced to below 0.5%. During storage, the final oxygen concentration will increase on the composition of the product. A high sugar and fat content will retard the diffusion process. After gas flushing, powder, especially for whole milk powder, must be filled in a multilayer (polyethylene inner) paper bags to prevent oxidation.

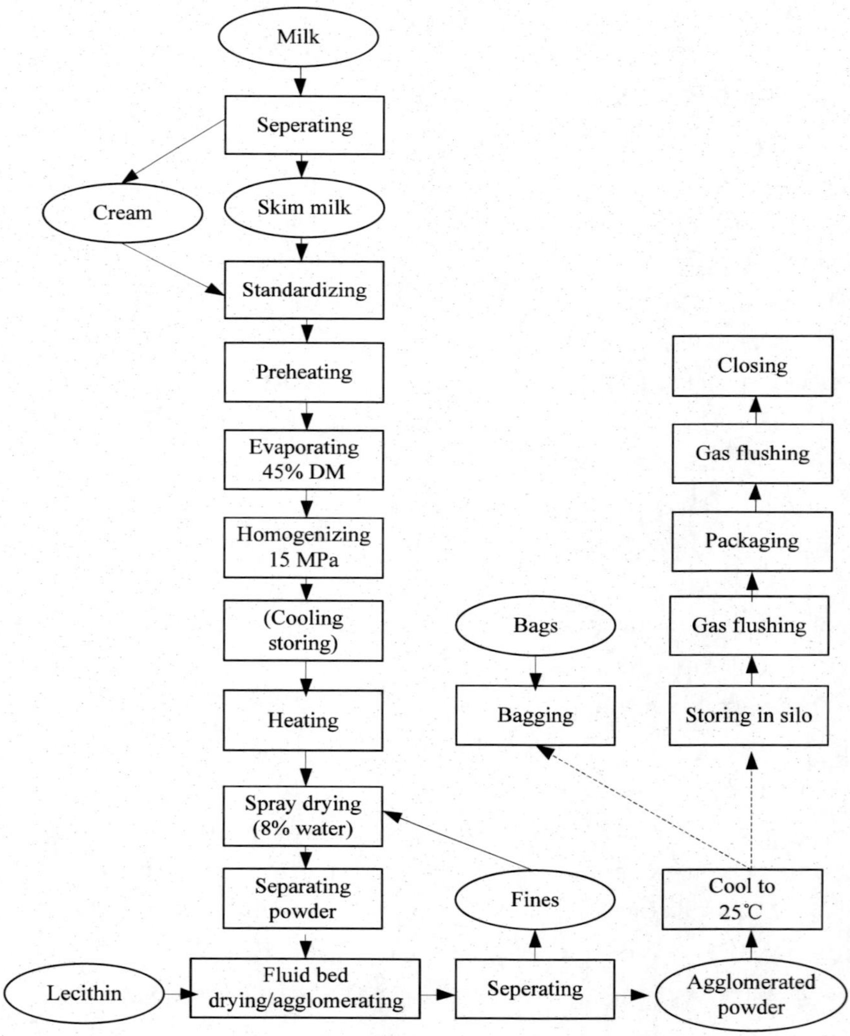

Figure 3. Example of the manufacture of whole milk powder.

Goat Milk Infant Formula

Breast milk is considered as the best food source for babies (Eidelman et al., 2012). Infant formula is commonly used to supplement breast milk when it is insufficient or unable to achieve breastfeeding for some reasons. Infant formula is widely considered to be the first choice for healthy bottle-fed infants. Milk proteins from skim milk and additional whey protein are added in the formula to improve the distribution of essential and semi-essential amino acids (Wit, 1998). As a breast milk substitute, infant formula is to simulate the nutritional composition and function of breast milk, so that non-breastfeeding infants are consistent with the growth and development of exclusively breast-fed infants.

Goat milk contains more whey proteins and bioactive factors than bovine milk, which may make it a good nutrition source for infants (Maathuis, Havenaar, He, & Bellmann, 2017). Consumer demand for goat milk infant formula has also been widely reported, using raw goat milk and homemade infant formula (Basnet, Schneider, Gazit, Mander, & Doctor, 2010; Ziegler et al., 2005). With regard to weight, length, head circumference, and BMI z-scores, there is no significant difference between infants fed by cow milk infant formula and goat milk formula. No differences between the groups in terms of motion consistency, crying duration, ease of settlement, or frequency of adverse events (Shao et al., 2014; Xu et al., 2015). Although goat milk is easily absorbed, infants who are allergic to milk protein cannot choose goat milk powder as a substitute. Cross-allergenicity between milk and goat milk proteins can lead to life-threatening complications (Ah-Leung et al., 2010; Pessler & Nejat, 2010).

Goat infant formula is being produced in several countries. With the widespread concern about the nutritional value of goat milk, goat milk formula has gradually become the development direction favored by many dairy enterprises, and various brands of goat milk infant formula have also been on the market, such as Kabrita and Karihome.

Manufacture Technology

Figure 4 is the flow chart of commercial production of goat milk based infant formula.

The production begins with raw milk pasteurization. In some companies, separating of raw milk into cream and skim milk goes first. The two streams are combined straight after the separator at a flow ratio which is based on the milk fat to protein ratio devised from the end product target. The ratio refers to the fat and protein required from milk. Fat and protein values used to derive this ratio exclude any fat and protein being added from other ingredients. This fat standardized milk is transferred to the standardized milk silo where the target volume equates to the total amount of milk used for each batch.

Table 1. Addition order and mixing time of nutrients

Order	Addition order	Mixing time
1	Calcium compounds	15 min
2	Vitamin premix	10 min
3	Mineral premix	20 min
4	Other nutrients (such as Choline chloride)	10 min

The main process is similar to whole milk powder production. Significant difference is nutrient addition. Vitamins and minerals should be added (Table 1). After one nutrient injected into the blender, it is necessary to wash ingredient container by water then injected into the blender. After that, the next ingredients were dissolved.

Critical Control Point and Control Measures

During the whole production, critical control point (CCP) is identified to reduce the risk. Different products have different control measures. Table 2 is an example of CCPs controlled in goat milk based infant formula production.

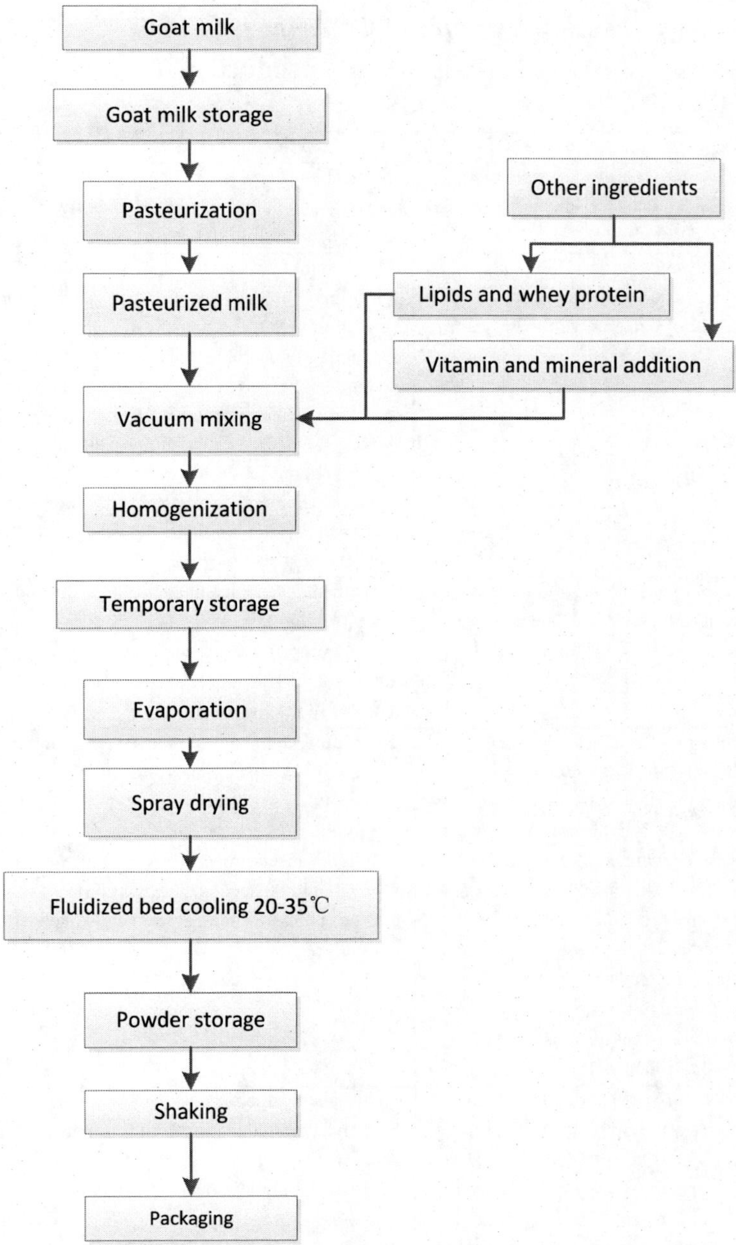

Figure 4. Manufacture of powdered goat milk infant formula.

Table 2. Example of CCPs for goat milk based infant formula production

CCPs	Control measures
Acceptance of raw milk: The raw milk entering the factory for each tank is sampled and tested.	Physical and chemical indicators, acidity, impurity and microbial indicators of fresh milk are tested to control the quality of fresh milk.
Raw milk pasteurization: This step pasteurizes the raw milk, and controls the total number of bacteria within 200 cfu/ml by controlling the temperature and time.	The pasteurization temperature parameter is set at 85°C, the allowable deviation is ± 2°C, and the heating time is 15 seconds; Preheating temperature is kept at 40-45°C; If the heating temperature is not up to the requirement, it must be reheated before it can be discharged. Among them, the heating temperature and time process should be strictly controlled, and the batch inspection should be recorded. When the process meets the requirements, it is allowed to enter the next section.
Vitamins and minerals addition: Calculate the amount of the nutrients and add them in order.	After the nutrients weighed, the quantity and weight should be double checked. The requirement of addition: confirm the addition after the review is completed; the monitor should be on.
Direct steam injection heating: the evaporation is a process of concentrating and heating the standardized mix by using an evaporator. When the concentration of the material flowing out of the return valve reaches 1040 kg/m^3, the discharge valve is opened and the return valve is closed. and the steam pressure, temperature and feed amount are further adjusted according to the discharge concentration, so that the discharge amount is controlled at 1.44-3.66 m^3/h.	The heating temperature must be reached during production and the temperature should be steady. If the heating temperature is not enough to supply the next process, the discharge valve should be closed immediately, and the return valve should be opened. The next process should not start unless the temperature adjustment is normal.
Spray drying: The evaporated material enters the high-pressure pump. It is sent to nozzle for spray drying. The pressurized material is sprayed at the top of the spray drying tower. The air inlet temperature is controlled at 170-195°C, and the outlet air temperature is 85-95°C. The concentrated milk droplets are dried into powder particles.	Stable steam supply Effective communication with the energy department Real-time monitoring of the inlet and outlet temperatures through the computer and the data should be recorded at the same time.

CCPs	Control measures
Powder shifting: Cut and open the inner bag with scissors, feed the powder into the vibrating screen. The powder enter silo through the 16-mesh vibrating screen.	Check the screen before starting every day Change the screen monthly
Metal foreign object detection: Three test pieces of different material sizes (iron: 1.0 mm; non-ferrous: 1.2 mm; stainless steel: 1.5 mm) were placed into the detection area of the metal detector. When the metal detector detects the test blocks, the yellow indicator light on the metal detector display will flash, and the reject device will remove the test objects. If all the three are removed at the same time, the metal detector is operating normally, and the metal detection can be performed normally. If the test objects are not properly removed, recalibrate the metal detector until all the objects are completely removed.	Test block (iron: 1.0 mm; non-ferrous: 1.2 mm; stainless steel: 1.5 mm); Perform three calibrations before production, at the end of production, and during the production process, and fill in the records.

GOAT MILK BUTTER

Butter is about 80% fat which is separated by mechanical centrifugation. Butter has a semi-hard and crystallized status, and it contains 80 to 85% fat and 15 to 20% water (Omo, 2003). Types of goat or sheep milk butter are similar to those of cow milk. However, due to the lack of agglutinin, the hardness and melting point of goat milk butter are significantly lower than cow milk butter, which makes the product quality is very hard to control in household or a factory setting (Omo, 2003). The most requirement properties of butter are below:

(i) Flavor: Nature fat flavor is encouraged, avoid lipolysis or volatile contaminants exists; (ii) Shelf life: For cultured-cream butter, avoiding molds, yeasts growth could improve the storage time and flavor. Lipases produced by psychrotrophs could cause lipolysis and further lead to off-flavor. Furthermore, during the low temperature storage, even at -20 °C, the autoxidation process of the fat also leads to a soapy rancid flavor; (iii)

Consistency: Because of the aggregation process of fat crystals, butter perform a high value of firmness. Consistency makes the butter a stable shape, no oiling off, and easily spreadable properties. However, consistency properties of butter could be influenced by the composition of fat and temperature; (iv) Color and homogeneity: Less problem has been found; (v) Yield: Butter yield is influenced by content of fat or water. For example, water content below 16% leads to a loss of yield; (vi) By-products: The most common by-product of butter is buttermilk. The worth of buttermilk is different in varies countries.

Manufacture

The skimming is essential to separate fat.

Pasteurization kills microorganisms, inactivates enzymes and makes the butter more resistant to oxidative deterioration. However, cooked and grassy flavors might be caused by heating treatment. In order to solve this problem, a vacreator has been used in cream pasteurization. It could remove off-flavor compounds from cream during the vacuum cooling process (Walstra et al., 2006).

Starter culture addition could improve aroma properties of butter. Study found that freeze-dried mesophilic aromatic lactic cultures addition could improve the content of lactic acid in butter making. Furthermore, some flavor components, primarily diacetyl, also can be produced by lactic culture (Rodríguez, Bunger, Castro, Sousa, & Empis, 2003). Fermented temperature and the concentration of lactic cultures can significantly improve the lactose content. However, the effect of fermentation time is not significant. Researchers reported that an optimal condition (concentration 8.8 U/100 L for 7 hours at temperature 28°C) in butter making and achieved a good sensory quality in appearance, flavor, texture, and overall quality of final product.

Ripening serves to sour the cream and to crystallize fat.

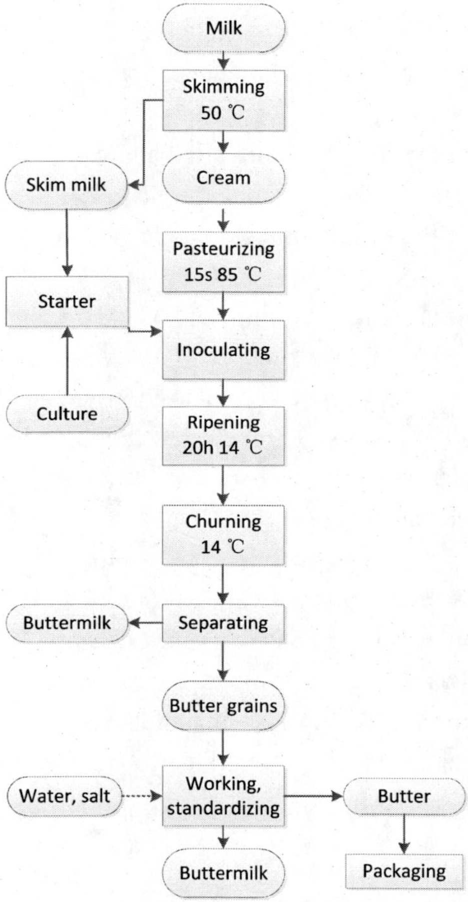

Figure 5. Traditional butter manufacture process. Adapted from (Walstra et al.2006).

The method of churning at low temperature can separate buttermilk and it can also reduce fat loss. During churning process, cream is added in a large vessel (also called dashboards) and rotated at low rotate speed for 20 minutes. After churning, butter grains have been collected and it need to be washed by water or butter milk. The washing temperature could significantly influence the retention rate of butter grains which has low firmness value. Meanwhile, the following drain off process is contributed to the butter moisture. In the butter making history, washing and draining are two important processes in quality control. But nowadays, washing is only used to control the temperature of butter. On all accounts, in order to form a correct firmness and particle size of butter grains, churning should be proceeded in a right way, and it can be described below:

1) Rotation mass should be converted and continued;
2) Disperse and standard the moisture in butter;
3) Incorporate the starter cultures, lactic acid and/or starter distillate.

It is worth noting that working and standardizing the butter could achieved by squeezing it through a special machine (for example, roller for falling height or perforated plate for subsection). Water content should be checked during this process. Sometimes, addition water in butter is necessary for standardization.

In Turkey, there is a special type of goat butter called "yayik butter" which was produced from yogurt. Due to the long shelf life and pleasant flavor, yayik butter is popular in Turkey. Traditional yayik butter can be made by two different ways, one is from fresh yogurt, and the other is from goat skin which has rennet to remove the whey (Nunez, 2016; Sağdıç, Dönmez, & Demirci, 2004). The manufacture process is completely different with the traditional butter making process mentioned above (Figure 5).

Properties of Goat Butter

Goat butter has been used for a long time like edible fat and source of energy in the African and Middle Eastern countries, especially in the rural areas. Goat butter has lower melting point than the cow butter, which gives it a soft consistency. As the goat butter does not contain carotene, it has a white color. Goat milk butter plays a significant role in the development of the organoleptic characteristics of certain foods (Nobuhiko, Shigeru, Yousuke, Hitoshi, & Hironori, 2005; Peterson & Reineccius, 2010).

The quality of butter is closely related to its physicochemical and microbiological characteristics. The quality of goat cream depends on several factors that are related to the quality and production conditions of the raw milk used for butter manufacture. Goat butter contains all the bacteria that may be contained in milk. In addition, lactic acid bacteria can also affect the sensory quality of butter.

Microbiological Properties

The average counts of total mesophilic bacteria are $2.07 \times 10^6 \pm 2.11$ cfu/g, which is high in all butter samples. This may be due to high numbers of microbes in raw goat milk, or due to an additional contamination during production. Coliform counts were not found in samples, which is an indicator of cleanliness of equipment and operation (Idoui, Habiba, & Nada, 2013).

Table 3. The microbial populations (CFU/g) in a traditional goat milk butter

Sample	1	2	3	4	5	6
Total bacteria (10^6)	1.3	0.57	1.07	6.76	1.27	1.45
Mould and yeast (10^2)	7	3	1	5.5	106	112
Total coliforms (10^4)	0	0	0	0	0	0
Lactic acid bacteria (10^5)	1.37	2.35	0.71	3.76	7.96	4.96
Staphylococci (10)	0	0	0	0	0	0
Salmonella	--	--	--	--	--	--
Total caseolytic bacteria (10^3)	4	51.5	1	556	612	298
Total lipolytic bacteria (10^3)	0	0	0	0	0	0
Total psychotrophic bacteria (10^6)	25	7	33.2	0.6	0.6	0.3

Note: Adapted from (Idoui et al.2013). --: Absence.

Fatty Acid Composition

The fatty acid profile of the 6 selected samples of a traditional goat milk butter are given in Table 4. Saturated fatty acids were identified and quantified. Among those, palmitic acid was the dominant one ranged between 20.5-50.6% followed by capric acid (20.3 - 40.0%), stearic acid (6.1

- 21.6%) and myristic acid (15.4 - 19.9%), lauric acid (7.7 -16.4%), caprylic acid (5.0 - 10.0%). These fatty acids have a relatively high melting point (above 44 C) and when they are present in great quantities in butter resulting in a firmer consistency (Idoui et al., 2013).

Table 4. Fatty acid profiles of traditional goat milk butter (peak area %)

Sample	1	2	3	4	5	6
Capric acid	-	-	-	39.55	27	20.26
Caprilic acid	-	-	-	9.98	6.67	5.04
Lauric acid	-	-	-	16.39	10.17	7.74
Linoleic acid	4.42	2.51	-	1.73	0.64	1.64
Margaric acid	1.23	-	-	1.83	2.22	1.53
Myristic acid	-	19.93	-	-	15.43	18.37
Oleic acid	19.49	18.42	-	4.35	9.31	16.5
Oxalic acid	-	-	-	0.92	-	0.16
Palmitic acid	50.56	42.03	-	-	20.45	-
Palmitoleic acid	-	-	-		0.29	0.34
Pentadecyclic acid	1.46	1.6	-	-	-	-
Stearic acid	20.7	13.44	-	21.61	6.08	19.29

Note: Adapted from (Idoui et al. 2013).

Fermented Goat Milk

As a transformation process working by microorganisms, fermented foods are not only worldwide popular but nutritious which can be dated back to 5000 BC.

According to the provision of FDA (Norberto et al., 2018), fermented milk, also called cultured milk, is defined as the "food produced by culturing one or more of the optional dairy ingredients (cream, milk, partially skimmed milk, and skim milk, used alone or in combination) with characterizing microbial organisms." Concentrated skim milk, nonfat dry milk, buttermilk, whey, or whey modified by partial or complete removal of lactose and/or minerals, may also allowed to be added to increase the nonfat solids content of the food. Furthermore, ingredients mentioned above shall be involved in the culturing process. All ingredients used should be safe for consumption.

People in Mediterranean countries have a long history of fermented goat milk production (Fisberg & Machado, 2015). Fermented goat milk products can be categorized according to the type of microorganism used during fermentation as follows.

1) Goat milk yogurt

   The type of microorganism used for yogurt fermentation contains streptococcus thermophilus and lactobacillus delbrueckii subsp. bulgaricus. Other subspecies of lactobacillus may also be used together with these two (FDA, 2018).

   There are two types of goat milk yogurt in the market, set and stirred, which are also named Dahi in India, has an appearance and mouth-feeling like pudding. It is because of the manufacturing process without destroying the curd after fermentation. Goat milk yogurt has a softer curd formation compared with cow and sheep milk yogurt. Studies on improving the texture of set goat milk yogurt have successfully solved the problem in different ways (Wang, Bao, Hendricks, & Guo, 2012). Stirred yogurt, which has a viscous and smooth taste, is made by stirring after fermentation. Food additive, such as fruits, jam, nuts or flavors can be added during the step of stirring. The diversification and marketability of stirred goat milk yogurt make it popular in many countries (Küçükçetin, Demir, Aşci, & Comak, 2011).

2) Goat milk kefir

   Traditionally, kefir was produced by the special starter named kefir grains, which contains Lactobacillus paracasei ssp. paracasei, lactobacillus acidophilus, lactobacillus delbrueckii ssp. bulgaricus, lactobacillus plantarum and lactobacillus kefiranofaciens (Zanirati et al., 2015). To avoid the defects of bulking or package leaking, some of the kefir like products were improved by using starter without yeast (Wang, Wang, Wang, & Guo, 2017).

3) Probiotic fermented goat milk

Probiotic microorganism can be used to produce probiotic fermented goat milk. It can also be combined with yogurt starter to make probiotic yogurt or with kefir starter to make probiotic kefir. Probiotic products, are the products that contain live organisms, found naturally in humans, are harmless and proven health benefits. The probiotics used in food production are mainly bacteria belong to *Lactobacillus* and *Bifidobacterium* (Pinto et al., 2017).

4) Labneh (concentrated yogurt)

Labneh is a kind of traditional dairy food in Lebanon. It is one of the main protein sources for the Lebanese people. It is produced by pouring the yogurt coagulum in a sanitized hanging cloth bag carefully, and then left the bag to drain the whey by gravity for about 12-18 hours at refrigeration temperatures until the desired total solids level is attained. The technology of centrifugation, recombination and ultrafiltration are used in modern Labneh manufacturing (Serhan, Mattar, & Debs, 2016).

5) Ayran (yogurt beverage)

Ayran is a kind of traditional non-sugary yogurt beverage in Turkey, which is famous for its salty taste and fermented sour aroma. It is a great thirst-quenched beverage with low viscosity and is very popular in summer months. As a region with high output of goat milk, Mediterranean countries have developed many kinds of fermented goat milk products.

After fermentation of non-sugary yogurt, stir and dilute the yogurt with potable water. During this process, 0.5-1% salt is added. Bubbles may be produced during stirring. After packaging the product is stored the Ayran at 4°C (Puniya, 2015).

Compare with Ayran made from cow milk, goat milk Ayran has higher intensity of goaty flavor, creamy taste and "throat burn" feeling. Using cow milk instead of part of the goat milk may reduce the goaty flavor (Uysal-Pala, Karagul-Yuceer, Pala, & Savas, 2006).

6) Frozen yogurt

Frozen yogurt, defined as a combination of yogurt and ice cream, has become a kind of modern and popular dairy food. It is also regarded as an ideal carrier of lactic acid bacteria, especially probiotics. Low storage temperature extends the shelf life of frozen yogurt products (Bezerra, Araujo, Santos, & Correia, 2015).

Goat Milk Yogurt

Goat is the first domesticated animal by human. The earliest yogurt in the world was believed to appear as far back as 10000 to 5000 BC in Mesopotamia, which was accidently made by goat milk with wild bacteria in goat skin bags (Fisberg & Machado, 2015). People simply boiled the goat milk and inoculated with yogurt then kept overnight at room temperature to make yogurt before yogurt starter discovered and industrialized (Tamime & Deeth, 1980).

Ingredients

Ingredients include raw goat milk, yogurt basic starter cultures, optional: dairy ingredients, sweeteners, stabilizers, probiotic cultures, prebiotic compounds, fat replacer, flavoring agents, color additives, vitamins.

Raw Goat Milk

Goat milk obtained from farm needs to do a microorganism counting, somatic cell counting and chemical analysis to guarantee the quality of the goat milk.

Yogurt Starter

Activate the cultures and prepare yogurt starter with skim milk or choose the ready-to-use yogurt starter cultures.

Dairy Ingredients

Since goat milk yogurt has a softer curd and lower firmness, goat milk powder, skim milk powder, casein powder, whey protein concentrate, whey protein isolate and some other dairy ingredients can be added into raw goat milk or low/non-fat goat milk before homogenization process of yogurt production to improve the texture of goat milk yogurt.

Sweeteners

Sweeteners such as sugar (sucrose), syrup, maltose or honey can be used in yogurt fermentation to improve the taste, cover the acid flavor and provide carbon source to the lactic acid bacteria. The optimal dose of sugar in yogurt fermentation is 5-8%.

Stabilizers

To reduce the syneresis and improve the gel strength of set-style goat milk yogurt and viscosity of stirred goat milk yogurt, plant/dairy based stabilizers are used before the process of fermentation of goat milk yogurt production (Lee & Lucey, 2010). Pectin products, mainly extracted from fruit peel, are often used as gelling agents in yogurt (Kazmierski, Wicker, & Corredig, 2003). Polymerized whey protein, induced by heating, acid, enzyme or other physiochemical methods from whey protein isolate, can also be used to improve gel formation. A cold-set gel forms when the pH goes down during fermentation (Wang et al., 2012).

Probiotics and Prebiotics

Probiotics are usually used together with traditional yogurt basic starter cultures to provide health benefits both in the gastrointestinal treat and immunity domains of human body (Pinto et al., 2017). Probiotic such as L. acidophilus can also be used independently to produce acidophilus goat milk. Prebiotics, such as galactooligosaccharide, fructooligosaccharide or inulin, are soluble dietary fiber that cannot be digested by human upper part of GI tract but can be used by probiotics. Yogurts containing both probiotics and prebiotics are called symbiotic yogurt (Walsh, Cheng, & Guo, 2014).

Fat Replacer

Fat replacer sometimes are needed to improve the texture of low-fat/non-fat goat milk yogurt. Dairy-based protein such as whey protein concentrate, microparticulated whey protein and modified tapioca starch, can be used alone or combined as a fat replacer in yogurt after physiochemical modification to meet the size of milk fat globule (Sandoval-Castilla, Lobato-Calleros, Aguirre-Mandujano, & Vernon-Carter, 2004; Zhang, McCarthy, Wang, Liu, & Guo, 2015).

Manufacture Technology

Filtration

In order to remove any solid particles that may become suspended, such as dirt, manure, straw, hair, clots and insects, a mechanical filter is used for the on-farm filtration (Vissers & Driehuis, 2009).

Homogenization

Homogenization is an important step in industrial production of whole-fat goat milk yogurt. It breaks down milk fat globules into smaller ones to promote the emulsification, which makes the texture uniform. As reported, homogenization could significantly improve the consistency and viscosity of fermented milk (Abrahamsen & Holmen, 1981). When skim milk powder, WPC, WPI, or other dairy/plant-based ingredients were added, homogenization can help to make them uniform distribution in yogurt (Vedamuthu, 1991). Homogenization may be not suitable for producing low-fat/non-fat goat milk yogurt because of the pore space in casein matrices in the yogurt. During whole-fat goat milk fermentation, fat globules can fill up the pores in casein matrices and increase the water-holding capacity of the yogurt. But when the fat is absent, the syneresis will be increased and a disadvantage to the texture of goat milk yogurt appears (Trachoo, 2002).

Pasteurization

Pasteurization is the most important step of ingredients preparation during yogurt production. Pasteurization and ultra high temperature (UHT) instantaneous sterilization are the two main methods to apply. Yogurt production, pasteurization usually be carried out at a temperature above 72°C, with a holding time of 15 seconds. It can destroy or inactivate organisms including bacteria, yeasts but not spores. UHT treatment is more efficient on raw milk sterilization. Preheat the milk or mixture to 95°C, and then treat it at a temperature of 142°C with a holding time of 3 seconds (Tobin, Fitzsimons, Kelly, & Fenelon, 2011). After fermentation, the growth of lactic acid bacteria can suppress activity of the harmful microorganism. Compared with UHT treatment, pasteurization is a better choice on goat milk yogurt production, due to the less loss in nutritive value, and energy conservation.

Heat Treatment

Goat milk mixture may also need to be heated at a temperature ranging from 80 to 95°C for 5 to 30 minutes to help induce whey protein denaturation and dissolve the thickening agent and prebiotics, like pectin and inulin (Wang, Gao, Zhang, Wang, & Guo, 2015). Denatured or polymerized of whey protein in the matrix improves the texture and consistency of the yogurt products. Adding polymerized whey protein to the milk mixture after pasteurization can also help to improve the firmness and the texture of the goat milk yogurt (Wang et al., 2012).

Inoculation

Pasteurized goat milk mix should be cooled to 42-45°C and inoculated with yogurt starter containing *L. delbrueckii* subsp. bulgaricus and *S. thermophilus*. Traditionally, yogurt starter was prepared by inoculating selected cultures from MRS broth to skim milk and fermented to let the cell grow (Li, Walsh, Gokavi, & Guo, 2012). Cultured skim milk used as yogurt starter will be inoculated (1%, v/v) to the cooled goat milk mixture. With the progress of technology on microorganism protection, the bacteria of yogurt starter culture have been freeze-dried or spray dried with preservative agent like protein or polysaccharide. The grain or powder of lactic acid bacteria with high level of activity can be used as a ready-to-use yogurt starter. It can be inoculated to the cooled goat milk mixture directly without preparation (Dave & Shah, 1997). *L. helveticus*, *L. acidophilus* and *Bifidobacterium* are generally considered as adjuncts in probiotic yogurt making.

Fermentation

The mix will be incubated at 43-45°C for 4-5 hours until the pH reaches 4.4-4.6. The formation of the curd and increase of the acidity indicates the formation of yogurt. The biochemical process of fermentation mainly including bacteria reproduction, acid coagulation, gel formation, and flavor substances generation. The optimum temperature at 43-45°C facilitates *L. delbrueckii* subsp. bulgaricus and *S. thermophilus* reproduce rapidly. The lactic acid bacteria ferment lactose and added sugars into lactic acid, and the pH of goat milk decreases. When the pH dropped to 4.6 (isoelectric point), casein-casein interactions occur and chains and three-dimensional network milk curd form (Nguyen, Afsar, & Day, 2018). Compared with cow milk, goat milk contains lower concentration of α_{s1}-casein, which results in poorer coagulation properties and curd firmness. During the process of fermentation, aroma compounds are the by-products by microorganisms, such as acetaldehyde, diacetyl, acetone, and 2-butanone, which are found very important in the formation of flavor of goat milk yogurt (Božanić, Tratnik, & Hruškar, 2003).

According to the technology of fermentation, yogurt mainly including two types, set-style yogurt and stirred yogurt. For set-style yogurt, cultured goat milk at 45°C will be filled in retail containers, such as plastic/glass cups, bottles or pots. After sealed, the goat milk mix will be fermented by lactic acid bacteria continuously in the retail containers. The gel structure will not be destroyed during and storage. For stirred yogurt, goat milk mix will be fermented in the large fermentation tanks. After gel formed, the curd will be disrupted by stirring to obtain a smooth and viscous texture. Flavoring additives such as fruits, jam, nuts or colorants can be added during step.

Cooling and Storage

After fermentation, yogurt products are cooled to 4°C to stop the work of microorganism and reduce further acid development. For set-style yogurt, it can be transferred to a cold storage room directly. For stirred yogurt, cooling process should be performed in the fermentation vat and then stirred smoothened and packaged (Lee & Lucey, 2010). Yogurt products with viable microorganism need to be stored at 4°C to ensure the survivability of the lactic acid bacteria. The viable bacteria count of the product shall not be lower than 10^6 CFU/mL during shelf life (Ravula & Shah, 1998). Yogurt product can also be sterilized after fermentation and it can be stored at room temperature. The advantages of sterilized yogurt are longer shelf life and easier for transportation.

Health Benefits of Fermented Goat Milk

Because of lower content of α_{s1}-casein, goat milk yogurt curd is softer and tender than that of cow milk. Similar to cow milk, caseins in goat milk can be divided into four types: κ-casein, α_{s1}-casein, α_{s2}-casein and β-casein. However, as the primary protein, content of α_{s1}-casein in cow milk ranges from 36-40% of total casein. While the content in goat milk is only 4-26%. Previous report considered that the structure properties of yogurt curd have been significantly influenced by the content of α_{s1}-casein. Low value of firmness in goat milk yogurt may be contributed to the low content of α_{s1}-casein (Clark & Sherbon, 2000). Although the low gel-strength texture may have some disadvantage for storage, transportation or consumer's preference. However, easily digestive of goat milk yogurt maybe due to the soft curd and have advantage for children or adult humans who suffer from gastrointestinal disturbances and ulcers (Kumar, Yadav, Kumar, Seth, & Kumar, 2016).

There are high contents of medium chain fatty acid or triglycerides found in goat milk which are easier for human especially children to digest and utilize. It is proved to have health benefits in malabsorption syndromes, chyluria, steatorrhea, hyperlipoproteinnemia, and in cases of intestinal resection, coronary bypass, premature infant feeding, childhood epilepsy, cystic brosis and gallstones. Medium chain triglycerides can also contribute to normal growth of infants (Bhattarai, 2012).

Another health benefit of goat milk is the low allergenic property. It is reported that about 2.5% of children before 3 years old has a cow milk allergy (Haenlein, 2004). Among the infants less than 3 months old fed by cow milk, the rate of occurring is as high as 12–30%. Goat milk was used to resolve this problem, about 30-40% of the treated children benefited from goat milk treatment. Goat milk has also been recommended to adult patients who are allergic to cow milk as a substitute. Between 40-100% of patients tolerate goat milk (Bhattarai, 2012). Reports indicated that α_{s1}-casein is the mainly factor for the allergenic reaction. Low content of α_{s1}-casein in goat milk maybe the explanation of why goat milk has a low allergenic reaction rate.

In addition, fermented goat milk is rich in vitamin A, vitamin B, vitamin K, niacin and affords larger amounts of minerals such as calcium, zinc and magnesium. Highly bioavailable phosphorus and iron absorption is an advantage for human health (Bergillos-Meca et al., 2015). Goat milk product is recommended as a valuable nutrition source for people especially for children and infants in many countries because of the better digestibility, healthier lipid metabolism and less allergenicity compared with other milks or dairy products.

Goat Milk *kefir*

Kefir, which was originated from Northern Caucasus and Eastern Europe, has been considered as an acidic-alcoholic fermented milk product with tart, carbonated and creamy flavor (Vahabzadeh & Ozpinar, 2018). It can also be called as *kef*, *kefy* or *keif* which was implied a pleasant flavor of taste in different countries. Traditional *kefir* is made by using commonly dairy substrate such as goat, sheep, buffalo or cow milk.

Chemical Properties of Kefir

Chemical composition of kefir varies widely. Milk composition, kefir grains composition and origination and fermentation or storage process can influence the chemical properties of kefir product. There are about 2.7% of protein (w/v), 0.6% of lactic acid (w/v) and less than 10% of fat (w/v) in kefir reported by Rosa et al. (2017). The value of pH in goat milk kefir is between 4.2 to 4.6. As the main metabolite in fermented milk, the content of lactic acid, CO_2 and ethanol in kefir is 0.8-1.0 (w/v), 0.08-0.2% (w/v) and 0.5-2.0% (w/v), respectively (Güven, Güven, & Gülmez, 2003). The profile of amino acid in kefir is similar to its fermentation substrate. High content of lysine (376 mg/100g), threonine (183 mg/100g), tryptophan (70 mg/100g), valine (220 mg/100g), methionine (137 mg/100g), phenylalanine (231 mg/100g) and isoleucine (262 mg/100g) have been found in cow milk kefir when compared with unfermented milk (Rosa et al., 2017). The content of lactose and lipid in kefir are also depended on the type of milk used in fermentation. Rich contents of vitamins (B1, B2, B5, C, A, K and carotene) have been found in kefir products (Liutkevičius & Šarkinas, 2004). Additionally, various minerals such as Ca, Zn, Cu, Mg, Mn, Fe, Co and Mo have been analyzed in goat milk kefir.

Manufacture Technology of Goat Milk *Kefir*

Traditionally, goat milk kefir has been manufactured by two different methods. One is "Russian method" (i.e., artisanal method) and another one is pure cultures method (i.e., commercial method) (Barukčić, Gracin, Režek, & Božanić, 2017). For Russian method (Figure 6), goat milk kefir was fermented by kefir grains on a large scale. After a series fermentation and percolate process, goat milk kefir was produced and stored at low temperature. Pure cultures method was employed from some pure cultures, which are predominant species in kefir grain, to make as direct-to-vat (DVS) starter culture by using freeze-drying technology. This starter shows good metabolic activity and fermentation efficiency, and it is convenient to use in kefir production (Bensmira, Nsabimana, & Jiang, 2010).

Milk Preparation

Kefir can be made by different fermentation substrates, such as mammalian milk (goat, cow, sheep, etc.) or plant milk. Tratnik, Bozanic, Herceg and Drgalic (2006) reported that the sensory score of kefir made by cow milk is higher than that of goat milk. de Lima et al. (2018) found that sheep's milk fermented by Brazilian kefir grains had high value of antioxidant and antimicrobial activities and high content of probiotic microorganisms. In addition to mammalian milk, plant milk can also be used for making kefir. It was reported that a soy milk kefir had high viable cells and acceptability and positive purchase intention (Liu, Chen, & Lin, 2002). However, the size of kefir grains is smaller when grown in soymilk.

Norberto et al. (2018) found that water-soluble soybean extract kefir showed consistent patterns of microbial, physiochemical and sensory properties. Cui, Chen, Wang & Han (2013) reported that walnut milk is good for kefir beverage making. Combined mammalian milk with plant milk was used to produce novel kefir product. Sabokbar and Khodaiyan (2015) evaluated the mixture of pomegranate juice and whey as a potential substrate for kefir making.

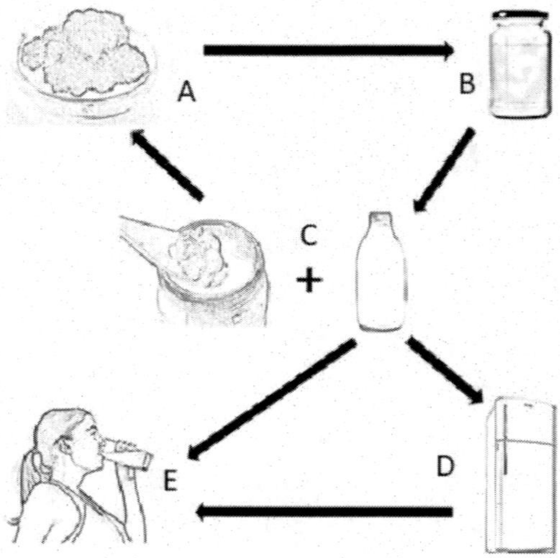

Figure 6. Russian method for goat milk *kefir* making. A. *Kefir* grains preparation. B. Addition of *kefir* grains to goat milk and fermentation. C. Filtration and separation of *kefir* grains. D. Stored at 4°C E. Adapted from (Rosa et al.2017).

Starter Cultures

Traditional kefir is produced by a starter culture named kefir grains (Figure 7). Kefir grains have an irregular mass, "pop-corn" or cauliflower shape, white or yellow color and gelatinous particles. The diameter of kefir grains ranges from 0.3-3.5 cm. The weight of kefir grains can be increased during fermentation process and this is why kefir grains can be re-used in kefir making. The chemical composition of kefir grains is different in various originations. An approximate 90% water (w/w), 3.2% protein (w/w), 0.3% lipids (w/w), 5.8% non-nitrogenous soluble substances (w/w) and 0.7% ash (w/w) have been found in kefir grains originated from Russia, Yugoslavia, Sweden and Bulgaria (Garrote, Abraham, & De Antoni, 2001).

Various species of microbial have been found in Kefir grains. These microbial species growth on the polysaccharide matrix (also called kefiran) can be divided into four groups: homofermentative, heterofermentative lactic acid bacteria, lactose and non-lactose assimilating yeast (Montanuci, Pimentel, Garcia, & Prudencio, 2012). Zanirati et al. (2015) reported that *lactobacillus paracasei* ssp. *paracasei*, *lactobacillus acidophilus*, *lactobacillus delbrueckii* ssp. *bulgaricus*, *lactobacillus plantarum* and *lactobacillus kefiranofaciens* are predominant species in *kefir* grains. In addition to these microbial species, *acetobacter aceti*, *acetobacter rasens* and a fungus *Geotrichum candidum* have also been found and isolated from *kefir* grains and *kefir* product (Zanirati et al., 2015). It is noteworthy that yeast is one of the characteristic microbial species in *kefir* grains. Diosma, Romanin, Rey-Burusco, Londero and Garrote (2014) found that almost 34 strains of yeast (15 strains of *S. cerevisiae*, 6 strains of *S. unisporus*, 4 strains of *I. occidentalis* and 9 strains of *K. marxianus*) in sugary kefir grains.

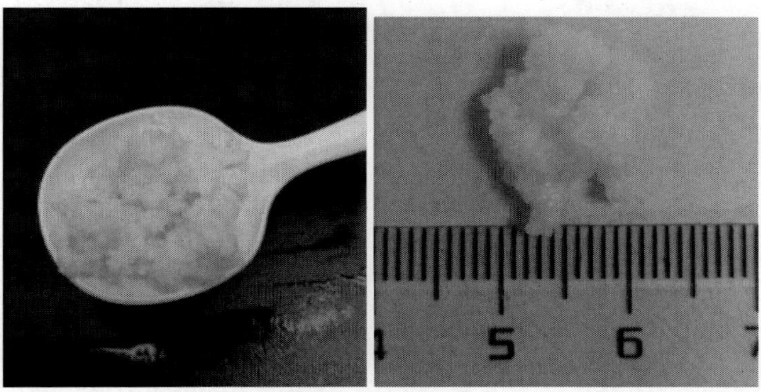

Figure 7. *Kefir* grains. Photo taken by H. Wang (2019).

Kefir products are not uniformed depending on the origin, storage or handling condition of kefir grains. It is difficult to obtain a quality and uniform manufactured kefir product by using kefir grains as starter culture. A lyophilized starter culture which has a mimic microbial composition with kefir grains has been used to produce commercial scale kefir. Commercial kefir starter cultures have been sold by many companies, such as Danisco Inc (Kefir D culture), Chr. Hansen Inc (Kefir 12 culture) or Institute Rossell Inc (Kefir Culture type C). But some obvious problems have been found in commercial kefir starter culture, such as short shelf life, poor kefir taste and aroma etc. (Nale, Tontul, Arslan, Nadeem, & Kucukcetin, 2018). Therefore, in order to solve the above problems, it is necessary to study the difference between traditional kefir grains and commercial kefir starter cultures.

Fermentation

Based on the metabolites produced, fermentation process can be divided into three types: lactic, yeast-lactic and mould-lactic fermentations. Goat milk Kefir is in the second category. Traditional kefir is produced from milk inoculation by a certain amount of kefir grains (1-3%) and 18-24 h long fermentation time at 20-25°C until the pH drops to 4.6. After incubation, kefir grains were removed by sieving and kefir product should be moved to 4-6°C in order to make secondary fermentation (Yildiz-Akgul, Yetisemiyen, Senel, & Yildirim, 2018). However, the ratio of *kefir* grains and fermentation substrate should range from 1:30 to 1:50 (w/v) and the optimal incubation temperature and time should be among 8-25°C and 10-40 h, respectively (Simova et al., 2002).

Quality Control

According to the standard of FAO/WHO (2011), goat milk kefir product should be made by kefir grains or starter culture prepared from kefir grains, lactobacillus kefiri, leuconostoc, lactococcus and acetobacter. Kefir grains should constitute both lactose fermenting yeasts and non-lactose-fermenting yeasts. The milk protein, titrable acidity, sum of microorganisms and yeasts should be up to 2.7% (w/v), 0.6% (w/v), 10^7 cfu/g, 10^4 cfu/g in goat milk kefir product, respectively. And the fat content should be lower than 10% (w/v).

There is no definite standard for goat milk kefir flavor, such as taste or aroma (Glibowski & Zielińska, 2015). Lactic acid, ethyl alcohol and flavor compounds in goat milk kefir make the volatiles differ from each other. A uniform flavor of traditional goat milk kefir was produced by the symbiotic activity of different microbiological composition in kefir grains. Muir, Tamime and Wszolek (1999) reported that the optimum flavor balance of kefir could be obtained when the ratio of diacetyl to acetaldehyde is 3:1. Addition of transglutaminase (m-TG) which creates a protein complex network in kefir could significantly enhance the sensory properties (Wroblewska, Kolakowski, Pawlikowska, Troszynska, & Kaliszewska,

2009). A commercial kefir product which has a controlled flavor has been produced through using of standard freeze-drying kefir starter culture. Meanwhile, a two-step fermentation technique was also developed to control the flavor of kefir (Yildiz-Akgul et al., 2018).

Health Benefits of Milk Kefir

Health benefits of goat milk kefir product are almost attributed to the various species of probiotic bacteria in kefir grains. Abundant probiotic bacteria correlated with health improvement. Like other fermented dairy products, goat milk kefir significantly reduced lactose content by 30% when compared with unfermented milk during fermentation (Rizk, Maalouf, & Baydoun, 2009). Antimicrobial properties which attributed to organic acids, H_2O_2, acetaldehyde and CO_2 during fermentation have been found in kefir product (Powell, 2006). Farnworth (2006) confirmed kefir grains had higher antibacterial activity than that of kefir products. Kefir grains can be used for preventing of infectious diseases and tumor development (Liu, Wang, Lin, & Lin, 2002). In addition, the microorganisms in kefir grains with their biological activities can be listed below: control of plasma glucose (Gomes, Bueno, de Souza, & Mota, 2014); anti-hypertensive (Quiros, Hernandez-Ledesma, Ramos, Amigo, & Recio, 2005); anti-inflammatory (Arthur & Jobin, 2013); antioxidative (Güven et al., 2003); anticarcinogenic (Hosono, Tanabe, & Otani, 1990); healing action (Lew & Liong, 2013). Another benefit of kefir has the capacity to lowering cholesterol level. Three ways for controlling the serum cholesterol by kefir were described by Prado et al. (2015). Firstly, bacterial cells in kefir grains could bind cholesterol and decrease the cholesterol absorption in the intestine. Secondly, among the increasing of production of short-chain fatty acids (SCFA) by kefir grains, the hydroxymethylglutaryl CoA (HMG-CoA) can be inhibited. Thus, the activity of cholesterol producing is decreased. The third, the bile salt hydrolase (BSH) could increase the bile acid deconjugation process in large intestine, resulting in the reduction of serum LDL-cholesterol concentrations in human intestine.

New Development of Kefir Product

Although kefir has been considered as a traditional probiotic beverage or fermented dairy product, it can be made as a novel product. Januario, Oliveira, Dias, Klososki and Pimentel (2018) developed a kefir ice cream flavored with fruit and honey by using kefir as base material. Plessas, Pherson, Bekatorou, Nigam and Koutinas (2005) made a bread by using kefir grains as baker yeast.

The effects of kefir grains on physical properties and structure of porous white and wheat bread were studied by Esteller, Zancanaro, Palmeira and da Silva Lannes (2006) and Filipcev, Simurina and Bodroza-Solarov (2007), respectively. A whey-cheese and a hard-type cheese were made using freeze dried kefir starter cultures (Dimitrellou, Kourkoutas, Banat, Marchant, & Koutinas, 2007; Katechaki, Panas, Rapti, Kandilogiannakis, & Koutinas, 2008). Lifeway co-orporation, one of the biggest kefir companies; manufactures kefir beverages, frozen products which are being sold all over the world. Other companies, such as Wallaby Yogurt Company, CocoKefir LLC also have their specialty kefir products which are popular among global consumers. In Russia, kefir consumption may help to reduce the risk of chronic diseases, hypertens ion, ischemic heart disease (IHD) and allergies (Figler et al., 2006).

Goat Milk Cheese

Like bovine milk, goat milk can be used to make a variety of different cheeses. Quality of goat milk cheese is influenced by many factors such as breeds, breeding program and milk composition. Goat breeds can affect cheese yield. Nubian and Alpine goats are two major dairy breeds in the United States. Milk from Nubian goats resulted in a much higher cheese yield than that from Alpine goats (Soryal, Beyene, Zeng, Bah, & Tesfai, 2005). Breeding program could also help improve the goat milk cheese composition, yield and fatty acid concentration. Crude protein and total solids of goat milk can be used to predict the yield of Chevre cheese (Guo, Park, Dixon, Gilmore, & Kindstedt, 2004).

Ripened low-fat cheese stored after 28 days had higher fracturability values, hardness, cohesiveness and masticability values than full fat cheese. Sensory properties of low-fat goat milk cheese are better than that of full fat goat milk cheese. 4-Methyl and 4-ethyl octanoic acids may be responsible for the waxy/animal flavor (Carunchiawhetstine, Karagul-Yuceer, Avsar, & Drake, 2010). Color of low-fat goat cheese was more yellow compared with full fat cheese which may be more attractive to consumers (Argueello, 2011).

Types of Goat Milk Cheese

There are many different types of goat milk cheese, such as Chevre cheese, Feta-style cheese, Cheddar cheese, mold cheese, etc. Flavor characteristics of these cheeses were significantly different. Chevre cheese tastes mild and had lower intensities for most attributes. Feta cheese was saltier, and Cheddar was mostly nutty, waxy and sweet. Mold-ripened cheese showed a musty-like and a goaty flavor which were more pungent and sharper than other cheeses (Talavera & Chambers, 2016). The majority types of goat milk cheese can be divided into three groups: a mild/blended group, a cooked milk/sharp/salty group, and a moldy/cheesy group (Talavera & Chambers, 2016).

Chevre Cheese

Chevre originated from French word “Shev”. “Shev” was originated from the Latin *capra*, which is the feminine version of goat. Chevre is used to characterize a cheese made from goat milk in French style. It is bland in flavor, creamy in body and texture and should be consumed fresh (Soryal et al., 2005). Chevre cheese (Figure 8) is one of popular soft cheeses on the market in United States and the main producers are Laura Chenel and

Vermont Creamery. It is a kind of fresh cheese and the shelf-life was relatively short (about 3 to 4 weeks). The manufacture process of Chevre cheese is shown in Figure 9. Briefly, the whole fat goat milk was pasteurized and then cooled at 22°C. Chevre starter and rennet was added to milk. After 16 hours' incubation process, the curd was broken up and addition of the salt. The salted curd was then pumped into nylon bags to drain the excess whey until 60% moisture level was reached. Finally, the cheese was packaged in a plastic tube. The relationship between Chevre yield and major components of goat milk was studied by Guo et al. (2004). Table 5 shows the relationship between Chevre yield and chemical components of the milk.

Figure 8. Chevre cheese. Photo taken by Guo (2019).

Table 5. Relationship between Chevre yield and chemical composition of goat milk

Equation	r^2	P-value
Y = 2.64 TS - 15.48	0.81	0.01
Y = 11.87 CN-13.30	0.75	0.01
Y = 8.61 CP – 12.76	0.79	0.01
Y = 3.85 F + 3.31	0.75	0.01
Y = 170.56 Ca – 8.46	0.64	0.01
Y = 2.31 F + 5.77 CN -5.97	0.80	0.01

CN: Casein, F: fat, CP: crude protein, TS: total solids, Ca: calcium. Adapted from Guo et al. (2004).

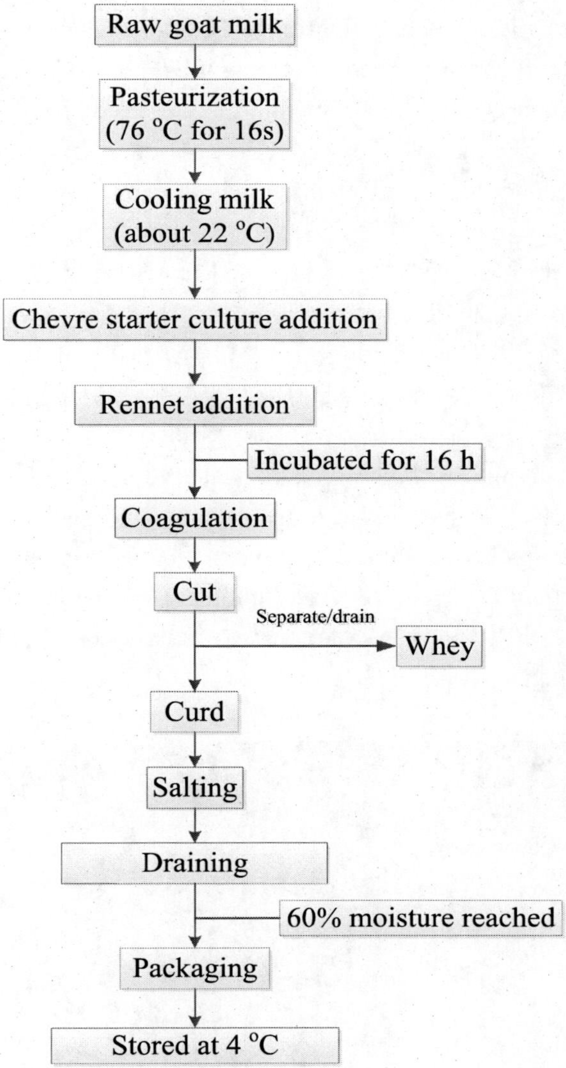

Figure 9. Flow chart of manufacture of Chevre (Guo et al., 2004).

Cheddar Cheese

Cheddar cheese (Goat milk Cheddar, see Figure 10) is a well-known and the most popular cheese in the world. Color of Cheddar cheese ranges from pale yellow to ivory white. During Cheddar cheesemaking process, coagulation of milk was achieved by adding the rennet enzyme (such as calf abomasum, papain enzyme or bromelain enzyme). Goat milk is a suitable raw material for Cheddar cheese production (Arlene, Kristijarti, & Ardelia, 2015). The Cheddar cheesemaking process is shown in Figure 11. Goat milk was pasteurized first.

Calcium chloride (0.2%, w/v) is added to the milk. Heated until the temperature reaches 30°C, starter culture (such as *lactobacillus lactic*) is added as much as 0.5% of the volume of milk. After 30 minutes, the enzyme is added, and this mixture is incubated for 30 minutes until the curd formation.

The curd is cut into 3 cm × 3 cm and heated to 38°C and then whey was separated. Salt (NaCl) is sprinkled in the curd and pressed at 1.4 atm for 12 hours at room temperature. Finally, the pressed curd is ripened in a ripening room for 1-12 months. Goat milk cheddar cheese was smooth in texture and has pale/white color if yellow colorant was not added due to it is lack of beta-carotene in goat milk.

Figure 10. Goat milk Cheddar cheese. Photo taken by Guo (2019).

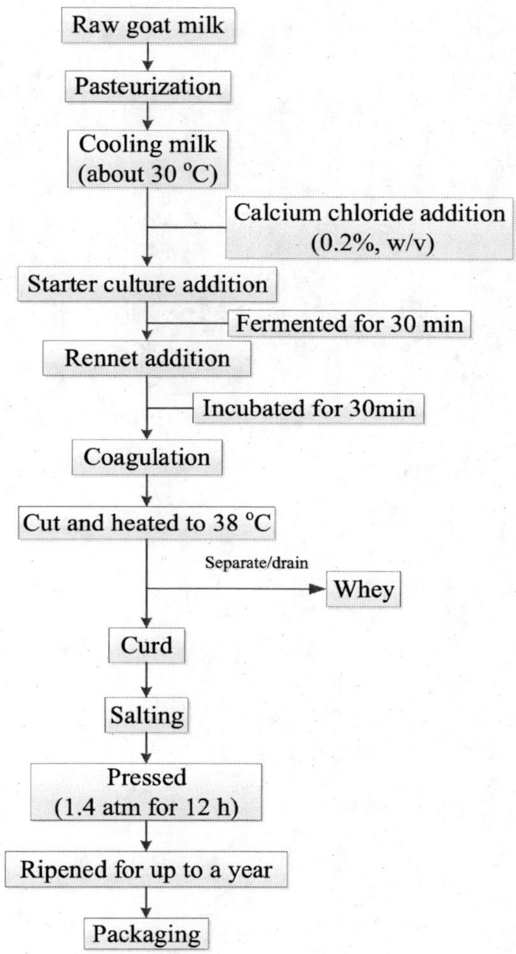

Figure 11. Flow chart of Cheddar cheese making from goat milk (Arlene et al., 2015).

Mozzarella Cheese

Goat milk Mozzarella cheese (Figure 12), original from Italy, is one of the most widely consumed cheeses in the world. The manufacture process of goat milk Mozzarella cheese is similar to Cheddar cheese, while the curd is kneading and stretching rather than pressed (Figure 13).

Figure 12. Goat milk Mozzarella cheese. Photo taken by Guo (2019).

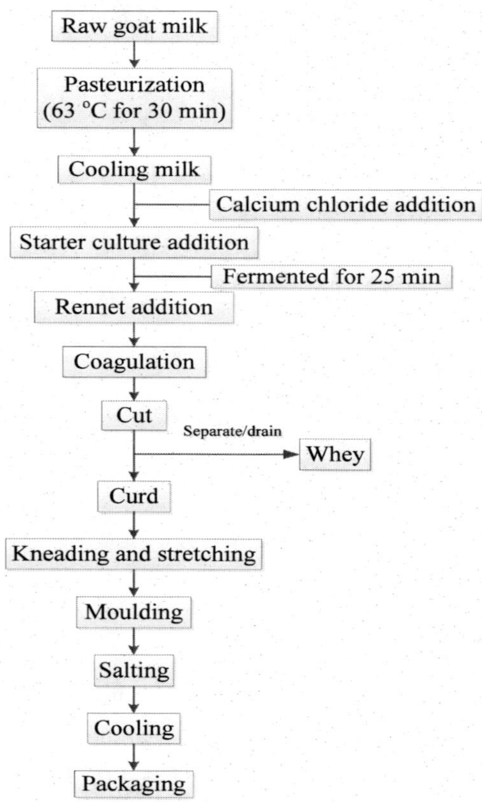

Figure 13. Flow chart of processing of goat milk Mozzarella cheese making (Paz et al., 2017).

The particular and plastic consistency structure of Mozzarella cheese is formed by a thermo-mechanical stretching process. This process results in the formation of casein fibers with fat and serum channels. The texture of Mozzarella cheese can be influenced by changing milk composition, whey pH and temperature, and storage conditions. Paz, Goncalvez de Oliveira, Villalva, Armada and Ramón (2017) found that the selection of pH at 5.3 could produce a suitable physicochemical, textural properties and microstructure of goat milk Mozzarella cheese.

CONCLUSION

Goat milk products are considered specialties in the United States. In this chapter the chemistry, manufacture technology liquid milk, goat milk powder, goat milk based infant formulas, butter, and major goat milk cheeses were discussed. The health benefits of fermented goat milk, yogurt, Kefir cheese and powdered products were reviewed. New developments in goat milk and new products processing technology were also summarized in this chapter.

REFERENCES

Abrahamsen, R. K., & Holmen, T. B. (1981). Goat's Milk Yoghurt Made From Non-Homogenized and Homogenized Milks, Concentrated By Different Methods. *Journal of Dairy Research*, 48(3), 457-463.

Ah-Leung, S., Bernard, H., Bidat, E., Paty, E., Rancé, F., Scheinmann, P., & Wal, J. M. (2010). Allergy to Goat and Sheep Milk without Allergy to Cow's Milk. *Allergy,* 61(11), 1358-1365.

Argueello, A. (2011). Trends in Goat Research, a Review. *Journal of Applied Animal Research, 3*9(4), 429-434.

Arlene, A., Kristijarti, A. P., & Ardelia, I. (2015). The Effects of The Types of Milk (Cow, Goat, Soya) and Enzymes (Rennet, Papain, Bromelain)

Toward Cheddar Cheese Production. *Makara Journal of Technology,* 19(1), 31-37.

Arthur, J. C., & Jobin, C. (2013). The Complex Interplay Between Inflammation, The Microbiota and Colorectal Cancer. *Gut Microbes,* 4(3), 253-258.

Barukčić, I., Gracin, L., Režek, J. A., & Božanić, R. (2017). Comparison of Chemical, Rheological and Sensory Properties of Kefir Produced By Kefir Grains and Commercial Kefir Starter. *Mljekarstvo,* 67(3), 169-176.

Basnet, S., Schneider, M., Gazit, A., Mander, G., & Doctor, A. (2010). Fresh Goat's Milk for Infants: Myths and Realities-A Review. *Pediatrics,* 125(4), E973-E977.

Bensmira, M., Nsabimana, C., & Jiang, B. (2010). Effects of Fermentation Conditions and Homogenization Pressure on The Rheological Properties of Kefir. *Lwt-Food Science and Technology*, 43(8), 1180-1184.

Bergillos-Meca, T., Cabrera-Vique, C., Artacho, R., Moreno-Montoro, M., Navarro-Alarcón, M., Olalla, M., & Ruiz-López, M. D. (2015). Does Lactobacillus Plantarum or Ultrafiltration Process Improve Ca, Mg, Zn and P Bioavailability from Fermented Goats' Milk? *Food Chemistry,* 187, 314-321.

Bezerra, M., Araujo, A., Santos, K., & Correia, R. (2015). Caprine Frozen Yoghurt Produced With Fresh and Spray Dried Jambolan Fruit Pulp (Eugenia Jambolana Lam) and Bifidobacterium Animalis Subsp. *Lactis Bi-07. Lwt-Food Science and Technology,* 62(2), 1099-1104.

Bhattarai, R. R. (2012). Importance of Goat Milk. *Journal of Food Science and Technology Nepal,* 7, 107-111.

Božanić, R., Tratnik, L., & Hruškar, M. (2003). Influence of Culture Activity on Aroma Components in Yoghurts Produced from Goat's and Cow's Milk. *Acta Alimentaria,* 32(2), 151-160.

Carunchiawhetstine, M. E., Karagul-Yuceer, Y., Avsar, Y. K., & Drake, M. A. (2010). Identification and Quantification of Character Aroma Components in Fresh Chevre-Style Goat Cheese. *Journal of Food Science,* 68(8), 2441-2447.

Chilliard, Y., Ferlay, A., Rouel, J., & Lamberet, G. (2003). A Review of Nutritional and Physiological Factors Affecting Goat Milk Lipid Synthesis and Lipolysis. *Journal of Dairy Science,* 86(5), 1751-1770.

Claeys, W. L., Cardoen, S., Daube, G., De Block, J., Dewettinck, K., Dierick, K., & Vandenplas, Y. (2013). Raw or Heated Cow Milk Consumption: Review of Risks and Benefits. *Food Control,* 31(1), 251-262.

Clark, S., & Sherbon, J. W. (2000). Alpha-S1-Casein, Milk Composition and Coagulation Properties of Goat Milk. *Small Ruminant Research,* 38(2), 123-134.

Cui, X. H., Chen, S. J., Wang, Y., & Han, J. R. (2013). Fermentation Conditions of Walnut Milk Beverage Inoculated With Kefir Grains. *LWT-Food Science and Technology,* 50(1), 349-352.

Dave, R. I., & Shah, N. P. (1997). Effectiveness of Ascorbic Acid As an Oxygen Scavenger in Improving Viability of Probiotic Bacteria in Yoghurts Made With Commercial Starter Cultures. *International Dairy Journal,* 7(6-7), 435-443.

De Lima, M. D. S. F., Da Silva, R. A., Da Silva, M. F., Da Silva, P. A. B., Costa, R. M. P. B., Teixeira, J. A. C., ... & Cavalcanti, M. T. H. (2018). Brazilian Kefir-Fermented Sheep's Milk, a Source of Antimicrobial and Antioxidant Peptides. *Probiotics and Antimicrobial Proteins,* 10(3), 446-455.

Dimitrellou, D., Kourkoutas, Y., Banat, I. M., Marchant, R., & Koutinas, A. A. (2007). Whey-Cheese Production Using Freeze-Dried Kefir Culture as a Starter. *Journal of Applied Microbiology,* 103(4), 1170-1183.

Diosma, G., Romanin, D. E., Rey-Burusco, M. F., Londero, A., & Garrote, G. L. (2014). Yeasts from Kefir Grains: Isolation, Identification, and Probiotic Characterization. *World Journal of Microbiology & Biotechnology,* 30(1), 43-53.

Eidelman, A. I., Schanler, R. J., Johnston, M., Landers, S., Noble, L., Szucs, K., & Viehmann, L. (2012). *Breastfeeding and The Use of Human Milk. Pediatrics,* 129(3), E827-E841.

Escareño, L., Salinas-González, H., Wurzinger, M., Iñiguez, L., Sölkner, J., & Meza-Herrera, C. (2012). Dairy Goat Production Systems. *Tropical Animal Health and Production,* 45(1), 17-34.

Esteller, M. S., Zancanaro, O., Palmeira, C. N. S., & Da Silva Lannes, S. C. (2006). The Effect of Kefir Addition on Microstructure Parameters and Physical Properties of Porous White Bread. *European Food Research and Technology,* 222(1-2), 26-31.

Fao. (2016). *Livestock Primary Production.*

Fao/Who. (2011). *Codex Standard for Fermented Milks.* In (Vol. Codex Stan 243-2003 (Revision 2008)).

Farnworth, E. R. (2006). Kefir–A Complex Probiotic. *Food Science and Technology Bulletin: Fu, 2*(1), 1-17.

FDA. (2018). Yogurt. In Federal Regulations (Vol. 21 Cfr 131.200). *USA: Department of Health and Human Services.*

Fedele, V., Rubino, R., Claps, S., Sepe, L., & Morone, G. (2005). Seasonal Evolution of Volatile Compounds Content and Aromatic Profile in Milk and Cheese from Grazing Goat. *Small Ruminant Research,* 59(2), 273-279.

Figler, M., Mózsik, G., Schaffer, B., Gasztonyi, B., Ács, P., Szili, B., ... & Szakály, S. (2006). Effect of Special Hungarian Probiotic Kefir on Faecal Microflora. *World Journal of Gastroenterology: WJG,* 12(7), 1129-1132.

Filipcev, B., Simurina, O., & Bodroza-Solarov, M. (2007). Effect of Native and Lyophilized Kefir Grains on Sensory and Physical Attributes of Wheat Bread. *Journal of Food Processing and Preservation,* 31(3), 367-377.

Fisberg, M., & Machado, R. (2015). History of Yogurt and Current Patterns of Consumption. *Nutrition Reviews,* 73(1), 4-7.

Fox, P., & Hoynes, M. (1976). Heat Stability Characteristics of Ovine, Caprine and Equine Milks. *Journal of Dairy Research,* 43(3), 433-442.

Garrote, G. L., Abraham, A. G., & De Antoni, G. L. (2001). Chemical and Microbiological Characterisation of Kefir Grains. *Journal of Dairy Research,* 68(4), 639-652.

Glibowski, P., & Zielińska, E. (2015). Physicochemical and Sensory Properties of Kefir Containing Inulin and Oligofructose. *International Journal of Dairy Technology,* 68(4), 602-607.

Gomes, A. C., Bueno, A. A., De Souza, R. G. M., & Mota, J. F. (2014). Gut Microbiota, Probiotics and Diabetes. *Nutrition Journal,* 13(1), 60.

Griffiths, M. W., & Phillips, J. D. (2010). Prediction of The Shelf-Life of Pasteurized Milk At Different Storage Temperatures. *Journal of Applied Bacteriology,* 65(4), 269-278.

Guo, M., Park, Y. W., Dixon, P. H., Gilmore, J. A., & Kindstedt, P. S. (2004). Relationship Between The Yield of Cheese (Chevre) and Chemical Composition of Goat Milk. *Small Ruminant Research,* 52(1-2), 103-107.

Güven, A., Güven, A., & Gülmez, M. (2003). The Effect of Kefir on The Activities of Gsh-Px, Gst, Cat, Gsh and Lpo Levels in Carbon Tetrachloride-Induced Mice Tissues. *Journal of Veterinary Medicine, Series* B, 50(8), 412-416.

Haenlein, G. (2004). Goat Milk in Human Nutrition. S*mall Ruminant Research,* 51(2), 155-163.

Horner, S. A., Wallen, S. E., & Caporaso, F. (1980). Sensory Aspects of Uht Milk Combined With Whole Pasteurized Milk. *Journal of Food Protection,* 43(1), 54-57.

Hosono, A., Tanabe, T., & Otani, H. (1990). Binding Properties of Lactic Acid Bacteria Isolated From Kefir Milk With Mutagenic Amino Acid Pyrolyzates. *Milchwissenschaft,* 45(10), 647-651.

Idoui, T., Habiba, R., & Nada, Z. (2013). Microbial Quality, Physicochemical Characteristics and Fatty Acid Composition of A Traditional Butter Made From Goat Milk. Annals. *Food Science and Technology,* 14(1), 108-114.

Januario, J. G. B., Oliveira, A. S., Dias, S. S., Klososki, S. J., & Pimentel, T. C. (2018). Kefir Ice Cream Flavored With Fruits and Sweetened With Honey: Physical and Chemical Characteristics and Acceptance. *International Food Research Journal,* 25(1), 179-187.

Katechaki, E., Panas, P., Rapti, K., Kandilogiannakis, L., & Koutinas, A. A. (2008). Production of Hard-Type Cheese Using Free or Immobilized

Freeze-Dried Kefir Cells As A Starter Culture. *Journal of Agricultural and Food Chemistry,* 56(13), 5316-5323.

Kazmierski, M., Wicker, L., & Corredig, M. (2003). Interactions of Β-Lactoglobulin and High-Methoxyl Pectins in Acidified Systems. *Journal of Food Science,* 68(5), 1673-1679.

Küçükçetin, A., Demir, M., Aşci, A., & Comak, E. (2011). Graininess and Roughness of Stirred Yoghurt Made with Goat's, Cow's Or A Mixture of Goat's and Cow's Milk. *Small Ruminant Research,* 96(2-3), 173-177.

Kumar, H., Yadav, D., Kumar, N., Seth, R., & Kumar, G., Arun. (2016). Nutritional and Nutraceutical Properties of Goat Milk-A Review. *Indian Journal of Dairy Science,* 69(5), 513-518.

Lee, W., & Lucey, J. (2010). Formation and Physical Properties of Yogurt. *Asian-Australasian Journal of Animal Sciences,* 23(9), 1127-1136.

Lew, L. C., & Liong, M. T. (2013). Bioactives from Probiotics for Dermal Health: Functions and Benefits. *Journal of Applied Microbiology,* 114(5), 1241-1253.

Li, S., Walsh, H., Gokavi, S., & Guo, M. (2012). Interactions Between Lactobacillus Acidophilus Strains and The Starter Cultures, Lactobacillus Bulgaricus and Streptococcus Thermophilus During Fermentation of Goats' Milk. *African Journal of Biotechnology,* 11(51), 11271-11279.

Liu, J. R., Chen, M. J., & Lin, C. W. (2002). Characterization of Polysaccharide and Volatile Compounds Produced By Kefir Grains Grown in Soymilk. *Journal of Food Science,* 67(1), 104-108.

Liu, J. R., Wang, S. Y., Lin, Y. Y., & Lin, C. W. (2002). Antitumor Activity of Milk Kefir and Soy Milk Kefir in Tumor-Bearing Mice. *Nutrition and Cancer,* 44(2), 183-187.

Liutkevičius, A., & Šarkinas, A. (2004). Studies on The Growth Conditions and Composition of Kefir Grains–As A Food and Forage Biomass. *Vet Zootec,* 25, 64-70.

Loewenstein, M., Speck, S., & Barnhart, H. (1979). Goat Milk and The Manufacture of Goat Milk Products. Paper Presented At *The Proc. 1st Ann. Florida Dairy Goat Prod. Conf.*

Loewenstein, M., Speck, S. J., Barnhart, H. M., & Frank, J. F. (1980). Research on Goat Milk Products: A Review1. *Journal of Dairy Science,* 63(10), 1631-1648.

Maathuis, A., Havenaar, R., He, T., & Bellmann, S. (2017). Protein Digestion and Quality of Goat and Cow Milk Infant Formula and Human Milk Under Simulated Infant Conditions. *Journal of Pediatric Gastroenterology and Nutrition,* 65(6), 661-666.

Mchenry, E. W. (1934). The Nutritional Value of Pasteurized Milk. *Canadian Public Health Journal,* 25(1), 22-25.

Montanuci, F. D., Pimentel, T. C., Garcia, S., & Prudencio, S. H. (2012). Effect of Starter Culture and Inulin Addition on Microbial Viability, Texture, and Chemical Characteristics of Whole or Skim Milk Kefir Efeito Do Tipo De Cultura Starter E Da Adição De Inulina Na Viabilidade Microbiana, Textura E Características Químic. *Ciência E Tecnologia De Alimentos,* 32(4), 580-865.

Muir, D. D., Tamime, A. Y., & Wszolek, M. (1999). Comparison of The Sensory Profiles of Kefir, Buttermilk and Yogurt. *International Journal of Dairy Technology,* 52(4), 129-134.

Nale, Z., Tontul, I., Arslan, A. A., Nadeem, H. S., & Kucukcetin, A. (2018). Microbial Viability, Physicochemical and Sensory Properties of Kefir Microcapsules Prepared Using Maltodextrin/Arabic Gum Mixes. *International Journal of Dairy Technology,* 71, 61-72.

Nguyen, H. T., Afsar, S., & Day, L. (2018). Differences in The Microstructure and Rheological Properties of Low-Fat Yoghurts From Goat, Sheep and Cow Milk. *Food Research International,* 108, 423-429.

Nobuhiko, I., Shigeru, W., Yousuke, Y., Hitoshi, T., & Hironori, N. (2005). Identification of Novel Decenoic Acids in Heated Butter. *Journal of The Agricultural Chemical Society of Japan,* 69(12), 2416-2420.

Norberto, A. P., Marmentini, R. P., Carvalho, P. H., Campagnollo, F. B., Takeda, H. H., Alberte, T. M., ... & Sant'ana, A. S. (2018). Impact of Partial and Total Replacement of Milk By Water-Soluble Soybean Extract on Fermentation and Growth Parameters of Kefir Microorganisms. *LWT-Food Science and Technology,* 93, 491-498.

Nunez, M. (2016). Existing Technologies in Non-Cow Milk Processing and Traditional Non-Cow Milk Products. In *Non-Bovine Milk and Milk Products* (Pp. 161-185): Elsevier.

Omo, O. (2003). Processed Food Products and Nutrient Composition of Goat Milk. *Pakistan Journal of Nutrition,* 2(2), 68-71.

Pandya, A. J., & Ghodke, K. M. (2007). Goat and Sheep Milk Products Other Than Cheeses and Yoghurt. *Small Ruminant Research,* 68(1), 193-206.

Paz, N. F., Goncalvez De Oliveira, E., Villalva, F. J., Armada, M., & Ramón, A. N. (2017). Effect of Ph at Drainage on The Physicochemical, Textural and Microstructural Characteristics of Mozzarella Cheese From Goat Milk. *Food Science and Technology,* 37(2), 193-201.

Pessler, F., & Nejat, M. (2010). Anaphylactic Reaction to Goat's Milk in A Cow's Milk-Allergic Infant. *Pediatric Allergy and Immunology,* 15(2), 183-185.

Peterson, D. G., & Reineccius, G. A. (2010). Determination of the Aroma Impact Compounds in Heated Sweet Cream Butter. *Flavour and Fragrance Journal,* 18(4), 320-324.

Petrus, R. R., Loiola, C. G., & Oliveira, C. A. (2010). Microbiological Shelf Life of Pasteurized Milk in Bottle and Pouch. *Journal of Food Science,* 75(1), M36-M40.

Pinto, J. M., Sousa, S., Rodrigues, D. M., Malcata, F. X., Duarte, A. C., Rocha-Santos, T. A., ... & Gomes, A. M. (2017). Effect of Probiotic Co-Cultures on Physico-Chemical and Biochemical Properties of Small Ruminants' Fermented Milk. *International Dairy Journal,* 72, 29-35.

Plessas, S., Pherson, L., Bekatorou, A., Nigam, P., & Koutinas, A. A. (2005). Bread Making Using Kefir Grains as Baker's Yeast. *Food Chemistry,* 93(4), 585-589.

Powell, J. E. (2006). *Bacteriocins and Bacteriocin Producers Present in Kefir and Kefir Grains.* Stellenbosch: University of Stellenbosch.

Prado, M. R., Blandon, L. M., Vandenberghe, L. P. S., Rodrigues, C., Castro, G. R., Thomaz-Soccol, V., & Soccol, C. R. (2015). Milk Kefir: Composition, Microbial Cultures, Biological Activities, and Related Products. *Frontiers in Microbiology,* 6(1177), 1-10.

Puniya, A. K. (2015). Traditional Fermented Dairy Products of Turkey. In *Fermented Milk and Dairy Products* (Pp. 430-455): Crc Press.

Quiros, A., Hernandez-Ledesma, B., Ramos, M., Amigo, L., & Recio, I. (2005). Angiotensin-Converting Enzyme Inhibitory Activity of Peptides Derived From Caprine Kefir. *Journal of Dairy Science,* 88(10), 3480-3487.

Ravula, R. R., & Shah, N. P. (1998). Selective Enumeration of Lactobacillus Casei from Yogurts and Fermented Milk Drinks. *Biotechnology Techniques,* 12(11), 819-822.

Remeuf, F. (1992). Physico-Chemical Properties of Goat Milk in Relation To Processing Characteristics. Paper Presented At *The Proceedings of The National Symposium on Dairy Goat Production and Marketing Oklahoma City Ok.*

Ribeiro, A. C., & Ribeiro, S. D. A. (2010). Specialty Products Made From Goat Milk. *Small Ruminant Research, 89*(2), 225-233.

Rizk, S., Maalouf, K., & Baydoun, E. (2009). The Antiproliferative Effect of Kefir Cell-Free Fraction on Hut-102 Malignant T Lymphocytes. *Clinical Lymphoma and Myeloma,* 9, S198-S203.

Rodríguez, A., Bunger, A., Castro, E., Sousa, I., & Empis, J. (2003). Development and Optimization of Cultured Goat Cream Butter. *Journal of The American Oil Chemists' Society,* 80(10), 9987-9992.

Rosa, D. D., Dias, M. M. S., Grzeskowiak, L. M., Reis, S. A., Conceicao, L. L., & Peluzio, M. (2017). Milk Kefir: Nutritional, Microbiological and Health Benefits. *Nutrition Research Review,* 30(1), 82-96.

Sabokbar, N., & Khodaiyan, F. (2015). Characterization of Pomegranate Juice and Whey Based Novel Beverage Fermented By Kefir Grains. *Journal of Food Science and Technology-Mysore,* 52(6), 3711-3718.

Sağdıç, O., Dönmez, M., & Demirci, M. (2004). Comparison of Characteristics and Fatty Acid Profiles of Traditional Turkish Yayik Butters Produced From Goats', Ewes' or Cows' Milk. *Food Control,* 15(6), 485-490.

Sandoval-Castilla, O., Lobato-Calleros, C., Aguirre-Mandujano, E., & Vernon-Carter, E. (2004). Microstructure and Texture of Yogurt As

Influenced By Fat Replacers. *International Dairy Journal,* 14(2), 151-159.

Serhan, M., Mattar, J., & Debs, L. (2016). Concentrated Yogurt (Labneh) Made of A Mixture of Goats' and Cows' Milk: Physicochemical, Microbiological and Sensory Analysis. *Small Ruminant Research,* 138, 46-52.

Shao, J., Zhou, Thomas, S., Gibson, R. A., Bo, L. N., Prosser, C. G., Lowry, D. J., & Maria, M. (2014). Nutritional Adequacy of Goat Milk Infant Formulas for Term Infants: A Double-Blind Randomised Controlled Trial. *British Journal of Nutrition,* 111(9), 1641-1651.

Shurpalekar, S. R., Chandrasekhara, M. R., Swaminathan, M., Sreenivasan, A., & Subrahmanyan, V. (1964). Studies on The Amino Acid Composition and Nutritive Value of The Proteins of Goat's Milk. *Indian Journal of Nutrition and Dietetics,* 25-27.

Simon, M., & Hansen, A. P. (2001). Effect of Various Dairy Packaging Materials on The Shelf Life and Flavor of Pasteurized Milk. *Journal of Dairy Science,* 84(4), 784-791.

Simova, E., Beshkova, D., Angelov, A., Hristozova, T., Frengova, G., & Spasov, Z. (2002). Lactic Acid Bacteria and Yeasts in Kefir Grains and Kefir Made From Them. *Journal of Industrial Microbiology and Biotechnology,* 28(1), 1-6.

Soryal, K., Beyene, F. A., Zeng, S., Bah, B., & Tesfai, K. (2005). Effect of Goat Breed and Milk Composition on Yield, Sensory Quality, Fatty Acid Concentration of Soft Cheese During Lactation. *Small Ruminant Research,* 58(3), 275-281.

Talavera, M., & Chambers, D. H. (2016). Flavor Lexicon and Characteristics of Artisan Goat Cheese from The United States. *Journal of Sensory Studies,* 31(6), 492-506.

Tamime, A., & Deeth, H. (1980). Yogurt: Technology and Biochemistry. *Journal of Food Protection,* 43(12), 939-977.

Tobin, J. T., Fitzsimons, S. M., Kelly, A. L., & Fenelon, M. A. (2011). The Effect of Native and Modified Konjac on The Physical Attributes of Pasteurized and Uht-Treated Skim Milk. *International Dairy Journal,* 21(10), 790-797.

Trachoo, N. (2002). Yogurt: The Fermented Milk. *Songklanakarin Journal of Science and Technology,* 24(4), 726-736.

Tratnik, L., Bozanic, R., Herceg, Z., & Drgalic, I. (2006). The Quality of Plain and Supplemented Kefir from Goat's and Cow's Milk. *International Journal of Dairy Technology,* 59(1), 40-46.

Uysal-Pala, C., Karagul-Yuceer, Y., Pala, A., & Savas, T. (2006). Sensory Properties of Drinkable Yogurt Made From Milk of Different Goat Breeds. *Journal of Sensory Studies,* 21(5), 520-533.

Vahabzadeh, S., & Ozpinar, H. (2018). Investigation of Some Biochemical Properties, Antimicrobial Activity and Antibiotic Resistances of Kefir Supernatants and Lactococcus Lactis Ssp Lactis Strains Isolated From Raw Cow Milk and Cheese Samples. *Kafkas Universitesi Veteriner Fakultesi Dergisi,* 24(3), 443-450.

Vedamuthu, E. (1991). The Yogurt Story: Past, Present and Future. Ii. *Dairy, Food and Environmental Sanitation (USA),* 11(5), 265-267.

Vissers, M., & Driehuis, F. (2009). On-Farm Hygienic Milk Production. *Milk Processing and Quality Management,* 1-22.

Walsh, H., Cheng, J., & Guo, M. (2014). Effects of Carbonation on Probiotic Survivability, Physicochemical, and Sensory Properties of Milk-Based Symbiotic Beverages. *Journal of Food Science,* 79(4), M604-M613.

Walstra, P., Wouters, J. T., & Geurts, T. J. (2006). *Dairy Science and Technology:* CRC Press.

Wang, C., Gao, F., Zhang, T., Wang, Y., & Guo, M. (2015). Physiochemical, Textural, Sensory Properties and Probiotic Survivability of Chinese Laosuan Nai (Protein-Fortified Set Yoghurt) Using Polymerised Whey Protein as A Co-Thickening Agent. *International Journal of Dairy Technology,* 68(2), 261-269.

Wang, H., Wang, C., Wang, M., & Guo, M. (2017). Chemical, Physiochemical, and Microstructural Properties, and Probiotic Survivability of Fermented Goat Milk Using Polymerized Whey Protein and Starter Culture Kefir Mild 01. *Journal of Food Science,* 82(11), 2650-2658.

Wang, W., Bao, Y., Hendricks, G. M., & Guo, M. (2012). Consistency, Microstructure and Probiotic Survivability of Goats' Milk Yoghurt

Using Polymerized Whey Protein as A Co-Thickening Agent. *International Dairy Journal,* 24(2), 113-119.

Wit, J. N. D. (1998). Nutritional and Functional Characteristics of Whey Proteins in Food Products. *Journal of Dairy Science,* 81(3), 597-608.

Wroblewska, B., Kolakowski, P., Pawlikowska, K., Troszynska, A., & Kaliszewska, A. (2009). Influence of The Addition of Transglutaminase on The Immunoreactivity of Milk Proteins and Sensory Quality of Kefir. *Food Hydrocolloids,* 23(8), 2434-2445.

Xu, M., Wang, Y., Dai, Z., Zhang, Y., Li, Y., & Wang, J. (2015). Comparison of Growth and Nutritional Status in Infants Receiving Goat Milk–Based Formula and Cow Milk–Based Formula: A Randomized, Double-Blind Study. *Food and Nutrition Research,* 59(1), 28613.

Yildiz-Akgul, F., Yetisemiyen, A., Senel, E., & Yildirim, Z. (2018). Microbiological, Physicochemical, and Sensory Characteristics of Kefir Produced By Secondary Fermentation. *Mljekarstvo,* 68(3), 201-213.

Young, O. A., Gupta, R. B., & Sadooghy-Saraby, S. (2012). Effects of Cyclodextrins on The Flavor of Goat Milk and Its Yogurt. *Journal of Food Science,* 77(2), S122-S127.

Zadow, J. G., Hardham, J. F., Kocak, H. R., & Mayes, J. J. (1983). The Stability of Goat's Milk to Uht Processing. *Australian Journal of Dairy Technology,* 38(1), 20-23.

Zanirati, D. F., Abatemarco, M., Sandes, S. H. D., Nicoli, J. R., Nunes, A. C., & Neumann, E. (2015). Selection of Lactic Acid Bacteria from Brazilian Kefir Grains for Potential Use As Starter Or Probiotic Cultures. *Anaerobe,* 32, 70-76.

Zhang, T., Mccarthy, J., Wang, G., Liu, Y., & Guo, M. (2015). Physiochemical Properties, Microstructure, and Probiotic Survivability of Nonfat Goats' Milk Yogurt Using Heat-Treated Whey Protein Concentrate as Fat Replacer. *Journal of Food Science*, 80(4), M788-M794.

Ziegler, D., Russell, S., Rozenberg, G., James, C., Trahair, T., & O'brien, T. (2005). Goats' Milk Quackery. *Journal of Paediatrics and Child Health,* 41(11), 569-571.

In: Goat Milk Chemistry …
Editor: Mingruo Guo
ISBN: 978-1-53617-299-7

Chapter 7

GOAT MILK PRODUCT QUALITY CONTROL

***Guorong Wang*[1] *and Mingruo Guo*[1,2,*]**
[1]Department of Nutrition and Food Sciences,
College of Agriculture and Life Sciences,
University of Vermont, Burlington, Vermont, US
[2]Key Laboratory of Dairy Science, Northeast Agricultural University,
Ministry of Education, Harbin, Heilongjiang, China

ABSTRACT

This chapter reviews the quality control of raw goat milk and the major goat milk products. Raw milk regulation of goat milk in many countries are similar to the bovine milk. The major difference between goat milk and bovine milk is the somatic cell count requirement. Summary of raw goat milk regulation in USA, Canada, China and EU have been illustrated in the chapter. Many factors could impact goat milk composition and quality, such as breed, dietary and milking process. Quality control of goat milk products (i.e., Fluid goat milk, goat milk powder, fermented goat milk, and goat milk cheese) is also introduced in this chapter.

Keywords: goat milk, quality, regulation, raw milk, fermented, powdered

[*] Corresponding Author's Email: mguo@uvm.edu.

INTRODUCTION

The regulations on goat milk is not that specific compared to its bovine milk counterpart. The majority of countries today are regulating goat milk the same as bovine milk. Goat milk production has its own characteristics that are different from cow milk, which makes the quality control slightly different from bovine milk products. This chapter will discuss current world wide regulations on goat milk and the quality control of different goat milk products.

RAW GOAT MILK REGULATIONS

Raw milk quality is the foundation of a successful production of goat milk and its products. Different countries may have different criteria to regulate the raw goat milk quality. In general, the quality raw milk should contain legal minimum level of all components, no objectionable flavor, be free of comtamination or adulterartion, and be safe to consume (Park & Guo 2006; Park, 2010; Garcia, Rovira, Boutoial, & Lopez, 2014). In this section, we will review the regulations of goat milk from major countries.

USA

Grade "A" Pasteruized Milk Ordiance (PMO) published by U.S. Department of Health and Human Services, Public Health Service, and Food and Drug Administration regulates all aspects of production, transportation and processing of all different milk products including goat milk. PMO defines goat milk as the normal lacteal secretion (free of colostrum) from healthy goat(s), and goat milk sold in retail package shall contain not less 2.5% milk fat and not less than 7.5% nonfat milk solids (FDA, 2017) compared to a minimum of 3.25% milk fat and 8.25% nonfat solids of cow milk defined by FDA 21 CFR Part 131 (FDA, 2018). A brief PMO standard of grade "A" raw goat milk for pasteurization was listed in Table 1. The requirement of grade "A" raw goat milk is almost the same as that of cow milk according to PMO except the somatic cell count (SCC). It is common that goat milk contains higher SCCs than bovine milk due to the different physiological process of milk secretion between goat (apocrine process) and cow (merocrine process) (Poutrel & Lerondelle, 1983; Boyazoglu & Morand-Fehr, 2001).

Table 1. Standard of Grade A raw goat milk in USA

Parameter	Standard
Temperature control	Cooled to below 10ºC within 4 hours of the first milking and to below 7°C within 2 hours after the completion of milking. The blend of temperature of the first milking and sebsequent milking does not exceed 10ºC.
Bacerial limits	Indivdual producer milk not to exceed 100,000 per ml prior to commingling with other producer milk. Not to exceed 300,000 per mil as commingled milk prior to pasteurization.
Drugs	No positive results on approved Beta lactam test methods for goat milk
Somatic cell count	Individual producer milk not exceed 1,500,000/mL

Adapted from Grade "A" PMO, 2017 Reversion by FDA, USA (FDA 2017).

Canada

In Canada, raw milk was regulated under the Canadian Food Inspection System (CFIS). The chemical and microbiological standard for raw goat milk is listed in Table 2.

Compared to FDA grade "A" PMO, the CFIS regulation has a tighter bacteria limits and same temperature and SCC limits. In addition, CFIS has freezing point limits while PMO does not. Freezing point testing is a rapid detection method to test raw milk quality, such as adulteration. Cow milk and goat milk have different freeze points (Harding, 1995). Heated milk has slightly higher freezing point than unheated raw milk (Janstova, Drackova, Navratilova, Hadra, & Vorlova, 2007).

Table 2. Canadian standard of raw goat milk

Parameter	Standard
Temperature	The first milking placed in the bulk milk tank shall be cooled to <=10°C within one hour, and further cooled to a temperature of >0°C and >=4°C within two hours after the completion of milking. When subsequent milking enter the tank, the blend temperature shall not rise >10°C
Bacteria count	Maximum 50,000 cfu/ml
Somatic cells	Maximum 1,500,000 somatic cells per ml
Veterinary drug residue	Negative for the presence of veterinary drug residues and inhibiory substance residues
Freezing point	Maximum: -0.554° Hortvet (-0.535°C) for goat's milk

Adapted from "National Dairy Code – Production and Processing Requirements, 7th edition (2015) by The Canadian Food Inspection System (CFIS) (CFIS, 2015).

China

Standards of raw milk in China is regulated under the National Standards System, known as the GB system. GB 19301-2010 is now the up to date version of regulation on raw milk. Goat milk standard is the same as cow milk except titratable acidity (TA) (Table 3). The TA for goat milk is 6-13°T and for cow milk is 12-18°T.

Table 3. China GB Standards of raw goat milk

Attributes	Limits
Protein (g/100g)	>=2.8
Fat (g/100g)	>=3.1
Impurity (mg/kg)	<=4.0
Nonfat solids (g/100g)	>=8.1
Acidity (°T) Cow milk Goat/sheep milk	12-18 6-13
SPC	<=200,000 cfu/ml

Adapted from China GB (19301-2010).

European Union

In Europe, the goat milk and its products are regulated under Commission Regulation (EC) No. 2073/2005 on microbiological quality for foodstuffs as milk and dairy products. The limits of SPC in raw milk EU is <100,000 no matter if it is individual or commingled milk (McLaughlin, 2006). After pasteurization, EU has a limit of Enterobacteriaceae <5 cfu/ml while many other countries do not have the limits for Enterobacteriaceae.

FACTORS THAT IMPACT GOAT MILK COMPOSITION AND QUALITY

Table 4. Factors affecting raw milk quality

Management and dietary factors
Proper feeding (i.e., balanced diet, amount of feeding, % of roughage feeding)
Proper handling of animals
Proper ventilation
Proper cleaning for barns and milking parlor
Prior and during milking
Feeding and handling prior and during milking
Use of recommended detergent, acid and sanizers
Cleanlinessof udder and teats, as well as milking worker
Use of properly functioning and clean equipment
Handling of the milk during and after milking
Attentiveness of dairy plant workers
Post milking and processing
Cooling
Transportation

Adapted from Park (2010).

The compostion of goat milk (mainly fat, protein, lactose and micro-components) are important to quality of milk and its products. For example, protein content (more precisely casein content) and fat content can impact the cheese firmness and yeild (Guo & Park, 2004; Fekadu et al., 2005; Raynal-Ljutovac, Gaborit, & Lauret, 2005). The goat milk compostion can be affected by breed, diet, stage of lactation, diet, environment, disease and other factors (Goetsch, Zeng, & Gipson, 2011; Garcia et al., 2014). The factors that affect milk quality before and after milking were shown in Table 4.

Breeds

There are six major dairy goat breeds and the milk production from greatest to the least are Alpine, Sannen, Oberhasli, LaMancha, Toggenhurg, and Nubian (Gipson & Grossman, 1989; Serradilla, 2001; Goetsch et al., 2011). Nubian has the least milk production but was the highest in fat among the six breeds (Haenlein, 1981). The varition in fat content was greater than the variation in protein content among different dairy goat breeds (Park, 2010).

Introducing high yield breed could potentially increase the milk yield. Serradilla (2001) reviewed the literatures of using high yield exotic breeds and concluded that 1) high yield exotic breeds may have better milk yields but lower protein, fat and solids content than the local breeds, and lower performances than in their region of origin; 2) seldom heterosis has been the reported, since the milk yield first crossed between exotic and local breeds are only the intermediate of their parents. Introduction of high yeild exotic breed should be based on long term and not only focusing the milk yield but also on the survival rates, disease resistance, reproductive performance (Serradilla, 2001). American Goat Society (AGS) uses a program called Dairy Herd Improvement (DHI) to improve the production of the purebred dairy goats through breeding, feeding and management decisions.

Feeding Management

Goat milk composition is also affected by diet, and could be manipulated to some extent by controling the feedstuff (Morand-Fehr & Sauvant, 1980; Morand-Fehr, Fedele, Decandia, & Le Frileux, 2007; Sanz Sampelayo, Chilliard, Schmidely, & Boza, 2007; Zervas & Tsiplakou, 2011). The farming and feeding management would significantly change the milk composition. A large goat milk proportion is being used to make cheese in Europe, thus the contents of fat and protein, and fat/protein ratio are critical quality control points for the goat cheese production.

It has been reported that reduced crude fiber level (high level of concentrate diets) in diet could decrease the fat level in goat milk (Morand-Fehr et al., 2007). Maintain a high fiber diet is critical to produce good quality of milk (Santini, Potchoiba, Fernandez, & Coleman, 1992; Pulina, 2008). For example, as the fiber increased from 14 to 26% on dry basis in diet, the goat milk fat creased from 2.48 to 3.32% (Santini et al., 1992). No matter if the dairy goat is of medium or low milk production, the correlations of dietary fiber and milk fat content are the same (Kawas, Lopes, Danelon, & Lu, 1991; Mele, Serra, Rafanelli, Conte, & Secchiari, 2005). The inclusion of fat supplements can also impact the goat milk yield (Chilliard & Ferlay, 2004).

The protein level can also be influenced by dietary nutrients, but mush less of variation has been observed compared to fat. Compared to dairy cow, the influence of dietary protein on the milk protein level is much less in dairy goat, as long as the dietary crude protein level is above the milk protein synthesis capacity (Sampelayo et al., 2007; Goetsch et al., 2011). Inclusion of protein concentration in diet does not change the protein level in milk but can increase the milk yield without compromise the protein level (Schmidely, 2002).

Appropriate incentive payment for goat milk in relation to milk composition is a widely used method to encourge milk producers to pursuit high quality raw milk by establishing bonuses and penalties (Pirisi, Lauret, & Dubeuf, 2007).

Bacteria and Somatic Cell Count (SCC) Controls

The bacteria and SCC count limits are mandatory regulation in raw milk. Temperature control is very important to keep low in bacteria count of raw milk. Cool the milk to below 4ºC as soon as possible (within two hours after milking) is required by US PMO (FDA 2017). Hygiene in milk handling is obvious to prevent microbial contamination and limit the bacteria counts (Delgado-Pertiñez et al., 2003).

SCC in goat milk is typically higher than cow milk due to the different physiological process of milk secretion between the two mammals. Somatic cells in goat had a distribution of 87% neutrophils, 9.9% macrophages, and 2.8% lymphocytes (Droke, Paape, & di Carlo, 1993). Maintain the health condition of lactating goats by limitating intra-mamary infection is very helpful to control SCC (Poutrel & de Cremoux, 1997).

Raw milk collected from individual producers or comingled should be cooled as soon as possible, and then filtered and stored. Filtration is used to remove the foreign particles; the simplest way is to filter milk through straining cloth. Mechnicala in-line milk filter is widely used in dairy industry.

Researchers have been testing the milk filter bacteria residue to monitor the pathogen presence in raw milk (Murphy et al., 2005). Centrifugal clarification can remove sediments and foreign materials as well as bacteria and spores (Gésan-Guiziou, 2010). It is reported that a new centrifugation called bactofugation can remove 86-92% SPC (Van den Berg, 1989), 94.1-98.7% spores (Van den Berg, 1989), and 95% somatic cells (Wieking, 2004).

MILK ADULTERATION

Intentional milk adulteration is a worldwide concern regarding to the safety and quality of milk. Milk adulterant ranges from nontoxic water and foreign sugar/protein/fat, to hazardous chemical like melamine and ammonium (Azad & Ahmed, 2016) Azad and Ahmed (2016) reviewed the common adulteration and rapid detection methods. With the advancement in rapid detection technology, milk adulteration can be closely monitored today.

Due to its seasonal production properties, goat milk could be adultered with cow milk. Given that some goat milk consumers may be allergic to cow milk, the adulterated milk could be very hazardous. Because of the similarity of goat and cow milk, detection of cow milk adulteration in goat milk is a challenge. A lot of researchers have been working on using different method to detect cow milk adulteration in goat milk by using the methods such as electronic tongue, HPLC, polymerase chain reaction and PCR assay (Bania, Ugorski, Polanowski, & Adamczyk, 2001; Chen, Chang, Chung, Lee, & Ling, 2004; Hurley, Coleman, Ireland, & Williams, 2004; Dias et al., 2009; Rodrigues et al., 2012).

HACCP for Goat Milk Products

The Hazard Analysis and Critical Control Points (HACC) program is a required in most of commercial industrialized goat milk plants to control the food safety problem of goat milk production. Each plant needs to develope their own HACCP plan according to the processing and conditions. In order to start a HACCP system, a HACCP plan and Sanitation Standard Operating Procedures (SSOPs) must be developed first. There are five preliminary steps for HACCP plan and seven HACCP Principles (Table 5).

Table 5. Five basic steps and seven principles for HACCP

Five Preliminary Steps for HACCP plan
1. Assemble HACCP team and bring together resouces 2. Describe the food and its method of distribution 3. Identify the intended use and consumers of the food 4. Develop a process flow diagram 5. Verify the operation diagrams
Seven HACCP principles
1. Conduct a hazard analysis 2. Identify critical control points 3. Establish critical limits for each critical control point 4. Establish monitoring procedures 5. Establish corrective actions 6. Establish record-keeping procedures 7. Establish verfication procedures

Fluid Milk Quality

Pasteurization is a critical step to kill pathogens in raw milk. PMO requires milk to be pasteurized within certain periods of time after it is collected. PMO requires the pasteurized milk has a SPC counts less than 20,000 cfu/ml and coliform less than 10 cfu/ml (FDA 2017). Figure 1 demonstrates a typical liquid milk processing flow chart. The liquid milk quality and its pasteurization are as important as the raw milk quality control, it does not only impact the fluid milk quality, but impact any other goat milk product made from the fluid milk such as goat milk powder, yogurt and cheese.

Before pasteurization, raw milk may subject to a separator to make nonfat or low-fat goat milk. Nutrients fortification such as vitamin D is required in some countries. Flavoring is also added in this step to make a variety of liquid goat milk products. After ingredients are added, the products will be subjected to a pasteurization. PMO defines couple several different combinations of temperature and time for liquid milk product (fat less 10%) (Table 6).

The combination of 145°F for 30min is vat pasteurization, all others are continuous pasteurization in Table 6. PMO has a very detailed regulation on the pasteurization such as monitoring temperature, flow rate (relate heating time), and diversion valve et al. to make sure all products are pasteurized above the minimum condition.

After milk is pasteurized, a step of homogenization is always followed. In-line homogenization is to decrease the fat globule from a mean value of 2.76 μm to less than 1 μm (Attaie & Richter, 2000). Homogenization does not significantly impact on bacterial counts, but it is very important to decrease or minimize the fat separation during shelf life (Lu, Wang, Li, & Zhang, 2013).

Currently the two-stage homogenization (2000psi for 1st stage, and 500psi for 2nd stage) at temperature of 50-70ºC is widely used in dairy plants. Fluid milk will be cooled down, filled and stored. The critical control point (CCP) is the pasteurization and time.

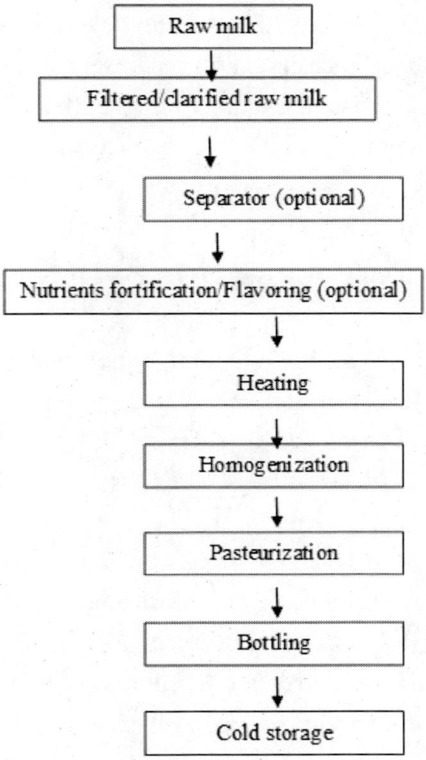

Figure 1. A flowchart of commercial fluid milk processing.

Table 6. Different milk pasteurization regime recognized by PMO

Temperature	Time
145°F (63°C)	30 min
161°F (72°C)	15 sec
191°F (89°C)	1 sec
194°F (90°C)	0.5 sec
201°F (94°C)	0.1 sec
204°F (96°C)	0.05 sec
212°F (100°C)	0.01 sec

Data source: PMO 2017.

Ultra-high temperature (UHT) is another heating process to make ambient shelf life stable products. UHT is processed at even higher temperature to achieve commercial sterilize for milk. For cow milk products, it is usually heat above 130ºC for 2-6 sec, however, goat milk is less stable compared with cow milk, sediment may present after UHT processing. Citrate and phosphate are usually added to goat milk to prevent the sediment in UHT goat milk (Boumpa, Tsioulpas, Grandison, & Lewis, 2008).

POWDERED GOAT MILK

The bacterial limits of milk powder are not exceed 10,000 per gram and coliform not exceed 10 per gram as regulated in PMO. Liquid goat milk is subjected to pasteurization and evaporation prior to drying. The pasteurization is a CCP to control the bacterial counts as discussed previously.

Evaporation is a step to concentrate goat milk to increase the drying efficiency. Vacuum evaporation is the most common technology in dairy industry by decrease the boiling point under vacuum to minimize the heat denaturation and off-flavor of product. Evaporation should be equipped with temperature and vacuum monitors as well as the solids content of post evaporation product. Cleaning routine should be implemented to prevent the buildup of thermo-philic bacteria especially for long production runs. If the evaporation is conducted by nanofiltration or other membrane technology, the routine monitoring system such as temperature control and bacteria build up should be in place. Membrane filtration is allowed to concentrate goat milk solids at even lower temperature to minimize the potential protein denaturation and nutrient loss.

An additional lactose crystallization between evaporation and drying is also necessary to control the quality of goat milk powder. Goat milk contains about 4.1% lactose, which is about 30% of the goat milk solids (Park 2017). Lactose has three forms (α-lactose, β-lactose, and lactose amorphous) at solids state, and only α-lactose monohydrate is a stable form, while the other two are very hygroscophic and not stable during storage (Gänzle, Haase, & Jelen, 2008). The crystallization step is allowed to let the super-saturated lactose in concentrate slowing transforms to the most stable α-lactose monohydrate to give an excellent flowablity of final powder and stable shelf life (Aguilar & Ziegler, 1994; Jouppila & Roos, 1994). The total solids, temperature and cooling curve of the evaporated concentrate should be optimized and monitored to maximize the α-lactose monohydrate crystallization.

Spray drying is the most common drying process for powder products. Inlet and outlet temperature should be closely monitored to prevent burnt on powder and control moisture level. A higher moisture in finished powder will cause shelf life issues. Different types of heater for the dryer, such as direct heater or indirect heater could impact the nitrite level in the finished powder (Manning, Coulter, & Jenness, 1968; Kelly, 1989). Nitrite is a chemical regulated by many countries especially for infant product applications.

Powder handling is the last step to ensure the quality and safety of powder (Fitzpatrick & Ahrné, 2005). Keeping packaging room clean and good personnel hygiene to prevent the introduction of contamination. Metal detector should be equipped before the package is sealed. The intact of package sealing should be checked. After powder is packaged, the condition of storage (such as temperature, humidity, free of strong smell/aroma materials) in warehouse and during transportation should be monitored.

FERMENTED GOAT MILK PRODUCTS QUALITY

Fermented goat milk can be roughly categorized into yogurt, Kefir, and cultured buttermilk based on the processing condition and cultures (Park 2010). The detailed information about fermented goat milk products were described in chapter 5. The process and quality control of fermented goat milk is very similar to its bovine milk counterpart; however, differences do exist between the two types of milk.

One of the most important chemical difference between goat milk and cow milk is the casein components. Goat milk contains much less α_{s1}- casein than cow milk (5.6% vs 38% of total casein) (Park 2017). This protein plays an important role in coagulation properties of goat milk (Ambrosoli, di Stasio, & Mazzocco, 1988). Lower α_{s1}- casein makes goat milk gel much weaker than cow milk (Ambrosoli et al., 1988; And & Guo, 2006). Various techniques have been studied to improve the fermented goat milk texture, such as fortification with milk powder (skim milk powder, whey protein, milk protein or casein products (Biliaderis, Khan, & Blank, 1992; And & Guo, 2006), concentration process (Marshall & El-Bagoury, 1986; Biliaderis et al., 1992), addition of stabilizers (Bruzantin, Daniel, da Silva, & Spoto, 2016), and enzymatic cross-linking (Farnsworth, Li, Hendricks, & Guo, 2006).

Listed in Table 8 is a brief summary and examples of fermented goat milk procedure and CCP. Heat treatment, inoculation, fermentation, cooling and packaging are the common CCP for most of fermented goat milk product.

Heat Treatment

A temperature of 85ºC for 30min (batch process) or 90-95ºC for 5-10min (continuous process) are typically used for fermented milk prior to inoculation (Wszolek, 2005). The purpose of heat treatment is not only to kill pathogen and other bacteria that may impact the fermentation but also to denature the protein (especially whey proteins) and induced whey protein-casein interactions to improve the resulting texture. That is why it has higher temperature and longer heating time compared to regular pasteurization.

Inoculation

Proper hygenic operations should be implemented to inoculation steps to avoid microbial contaminations. The inoculation dosage should strictly follow the culture direction.

Fermentation

The hygenic condition of fermentation tank or container should be verified to avoid microbial contamination. The temperature of fermentation should be closely controlled and monitored. The typical fermentation temperature for yogurt, kefir and buttermilk are 42-45°C, 22°C and 22°C, respectively (Table 7).

Cooling

The end of fermentation should be closely monitored. In order to obtain the best quality of fermented goat milk products. Fermentation should be stopped by immediately cooling to 4°C (except kefir which has a ripening process) after the pH or titratable acidity% reach to its target (Table 7). Stop the fermentation at the right pH is not only necessary to obtain the best flavor/texture, but also impact the shelf life microbiological stability.

Table 7. Manufacturing conditions and CCP of goat milk yogurt, kefir and buttermilk

Products	Culture	Incubation condition	Stop incubation at pH	An example of procedure with CCP
Yogurt	S. thermophilus L. bulgaricus	45.6°C for 5-6 hours	4.2-4.5	Raw milk -> standizaion/fortification (optional) -> homogenization -> heat treatment * -> innoculation *-> fermentation * -> cooling * -> packaging* -> storage*
Kefir	S. kefir T. kefir L. caucasius S. lactis	22°C followed by 10°C, 12 hours	4.5	Raw milk -> homogenization -> heat treatment* -> innoculation *-> fermentation* -> ripening -> cooling *-> packaging * -> storage*
Cultured Buttermilk	S. lactis S. cremoris L. citrovorum S. diacetylactis	22°C for 12-16 hours	4.5	Buttermilk/skim or whole milk -> heat treatment * ->innoculation * -> fermentation *->cooling* -> packaging *-> storage*

* indicates the critical control points.

Packaging and Storage

The containers and packaging machine should be sanitized properly, and packaged products should be stored below 4ºC in warehouse, transportation, and in stores before customer consumption.

GOAT MILK CHEESE QUALITY CONTROL

Both unpasteurized and pasteurized goat milk can be used to make goat milk cheese. Unpasteurized milk cheese, also referred as raw milk cheese, has a much richer flavor compared to the pasteurized milk cheese, but potential risks exist (Beuvier, 2004). Unpasteurized milk cheese has to be aged for a minimum of 60 days to be considered safe to customers (Peter 2000; FDA 2016). In Europe, raw milk is referred to the milk has not been heated beyond 40ºC (Peter 2000). Please refer to previous sections in this chapter for raw milk quality control.

Similar to bovine milk cheese, goat cheese can be categorized into soft, semi-soft, semi-hard, and hard cheese based on the firmness and moisture (Varnam and Sutherland 2001). Figure 2 demonstrated the simplified procedures of various cheese product. Raw milk should be collected, transferred and pasteurized accordingly. Typically, it is not necessary to homogenize the milk for cheese making because the homogenized smaller fat globule may impact the texture and flavor the finishing cheese (Nair, Mistry, & Oommen, 2000). Pasteurization is a critical control point, have a good recording keeping and CIP is very critical. Over pasteurization may induce the casein-whey interactions, thus hindering the renneting process (Ustunol & Brown, 1985).

Starters is always added to acidify the milk to facilitate renneting. When rennet is added, the coagulum is formed and then whey is drained, soft cheese is produced (non-ripened or reopened). For semi-hard or hard cheese, the rennet coagulum will be cut and whey will be drained, and also followed by dry mixing ingredients, mould, and ripened.

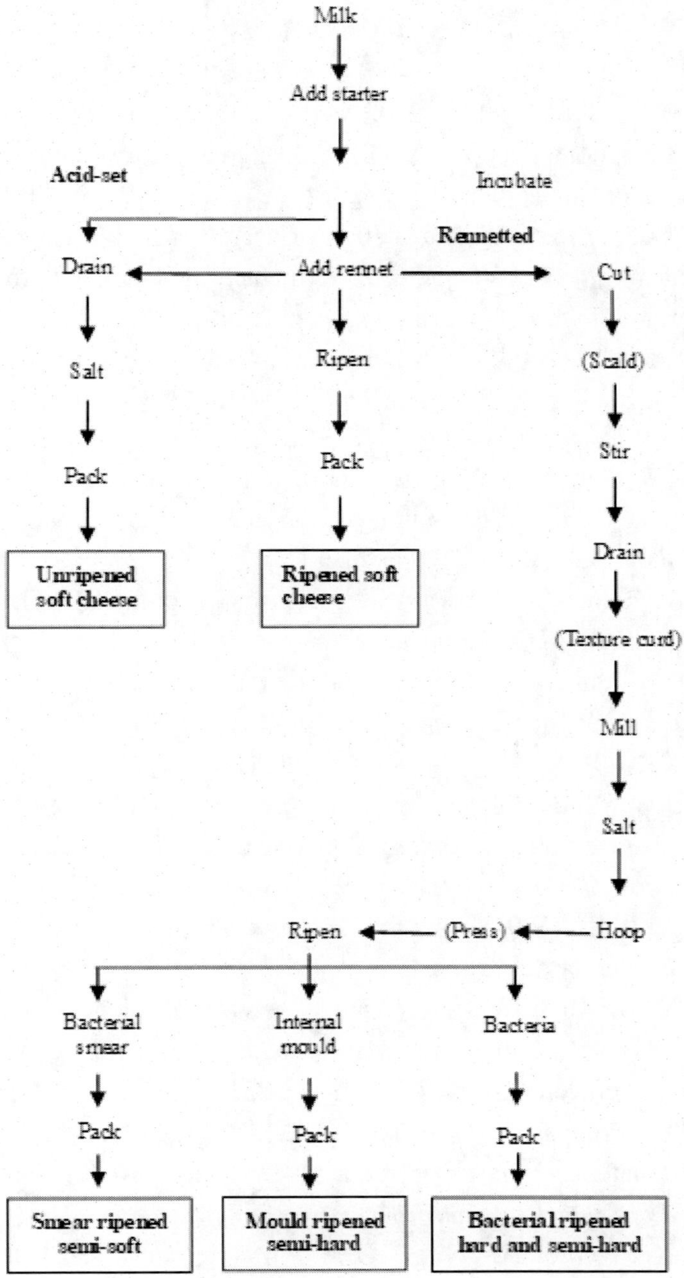

Figure 2. Simplified procedure for the manufacture of various types of cheese. Adapted from Varnam & Sutherland 2001.

REFERENCES

Aguilar, C. A., & Ziegler, G. R. (1994). Physical and microscopic characterization of dry whole milk with altered lactose content. 2. Effect of lactose crystallization. *Journal of Dairy Science*, 77(5), 1198-1204.

Alm, L. (1982). Effect of fermentation on lactose, glucose, and galactose content in milk and suitability of fermented milk products for lactose intolerant individuals. *Journal of Dairy Science*, 65(3), 346-352.

Ambrosoli, R., di Stasio, L., & Mazzocco, P. (1988). Content of αs1-Casein and Coagulation Properties in Goat Milk. *Journal of Dairy Science*, 71(1), 24-28.

And, J. L., & Guo, M. (2006). Effects of Polymerized Whey Proteins on Consistency and Water-holding Properties of Goat's Milk Yogurt. *Journal of Food Science*, 71(1), C34-C38.

Attaie, R., & Richter, R. L. (2000). Size Distribution of Fat Globules in Goat Milk. *Journal of Dairy Science*, 83(5), 940-944.

Azad, T., & Ahmed, S. (2016). Common milk adulteration and their detection techniques. *International Journal of Food Contamination*, 3(1), 22.

Bania, J., Ugorski, M., Polanowski, A., & Adamczyk, E. (2001). Application of polymerase chain reaction for detection of goats' milk adulteration by milk of cow. *Journal of Dairy Research*, 68(2), 333-336.

Beuvier, E. a. S. B. (2004). Raw milk cheeses. *Cheese: Chemistry, Physics and Microbiology*, 1, 319-345.

Biliaderis, C. G., Khan, M. M., & Blank, G. (1992). Rheological and sensory properties of yogurt from skim milk and ultrafiltered retentates. *International Dairy Journal*, 2(5), 311-323.

Boumpa, T., Tsioulpas, A., Grandison, A. S., & Lewis, M. J. (2008). Effects of phosphates and citrate on sediment formation in UHT goats' milk. *Journal of Dairy Research*, 75(2), 160-166.

Boyazoglu, J., & Morand-Fehr, P. (2001). Mediterranean dairy sheep and goat products and their quality: A critical review. *Small Ruminant Research*, 40(1), 1-11.

Bruzantin, F. P., Daniel, J. L. P., da Silva, P. P. M., & Spoto, M. H. F. (2016). Physicochemical and sensory characteristics of fat-free goat milk yogurt with added stabilizers and skim milk powder fortification. *Journal of Dairy Science*, 99(5), 3316-3324. *Production and Processing requirements*, 7th edition (Part I), (2015).

Chen, R. K., Chang, L. W., Chung, Y. Y., Lee, M. H., & Ling, Y. C. (2004). Quantification of cow milk adulteration in goat milk using high-performance liquid chromatography with electrospray ionization mass spectrometry. *Rapid Communications in Mass Spectrometry*, 18(10), 1167-1171.

Chilliard, Y., & Ferlay, A. (2004). Dietary lipids and forages interactions on cow and goat milk fatty acid composition and sensory properties. *Reproduction, Nutrition, Development*, 44(5), 467-492.

Delgado-Pertiñez, M., Alcalde, M. J., Guzmán-Guerrero, J. L., Castel, J. M., Mena, Y., & Caravaca, F. (2003). Effect of hygiene-sanitary management on goat milk quality in semi-extensive systems in Spain. *Small Ruminant Research*, 47(1), 51-61.

Dias, L. A., Peres, A. M., Veloso, A. C. A., Reis, F. S., Vilas-Boas, M., & Machado, A. A. S. C. (2009). An electronic tongue taste evaluation: Identification of goat milk adulteration with bovine milk. *Sensors and Actuators B: Chemical*, 136(1), 209-217.

Droke, E. A., Paape, M. J., & Di Carlo, A. L. (1993). Prevalence of High Somatic Cell Counts in Bulk Tank Goat Milk. *Journal of Dairy Science*, 76(4), 1035-1039.

Farnsworth, J. P., Li, J., Hendricks, G. M., & Guo, M. R. (2006). Effects of transglutaminase treatment on functional properties and probiotic culture survivability of goat milk yogurt. *Small Ruminant Research*, 65(1), 113-121.

FDA (2016). FY 2014–2016, Raw Milk Cheese Aged 60 Days. *Summary Report: Microbiological Sampling Assignment*.

FDA (2017). *Grade "A" pasteurized milk ordinance*, 2017 revision. U.S. Dept. of Health and Human Services, Public Health Service, Food and Drug Administration, 2017.

FDA (2018), *Part 131 – Milk and Cream, Tile 21-Food and Drug* (21CFR131.110).

Fekadu, B., Soryal, K., Zeng, S., Hekken, D. V., Bah, B., & Villaquiran, M. (2005). Changes in goat milk composition during lactation and their effect on yield and quality of hard and semi-hard cheeses. *Small Ruminant Research*, 59(1), 55-63.

Fitzpatrick, J. J., & Ahrné, L. (2005). Food powder handling and processing: Industry problems, knowledge barriers and research opportunities. *Chemical Engineering and Processing: Process Intensification*, 44(2), 209-214.

Gänzle, M. G., Haase, G., & Jelen, P. (2008). Lactose: Crystallization, hydrolysis and value-added derivatives. *International Dairy Journal*, 18(7), 685-694.

Garcia, V., Rovira, S., Boutoial, K., & Lopez, M. B. (2014). Improvements in goat milk quality: A review. *Small Ruminant Research*, 121(1), 51-57.

Gésan-Guiziou, G. (2010). 15 – *Removal of bacteria, spores and somatic cells from milk by centrifugation and microfiltration techniques* Elsevier, 349-372.

Gipson, T. A., & Grossman, M. (1989). Diphasic Analysis of Lactation Curves in Dairy Goats1. *Journal of Dairy Science*, 72(4), 1035-1044.

Goetsch, A. L., Zeng, S. S., & Gipson, T. A. (2011). Factors affecting goat milk production and quality. *Small Ruminant Research*, 101(1-3), 55-63.

Haenlein, G. F. W. (1981). Dairy Goat Industry of the United States. *Journal of Dairy Science*, 64(6), 1288-1304.

Harding, F. (1995). Adulteration of milk. *Milk Quality*. Springe, 60-74.

Hurley, I. P., Coleman, R. C., Ireland, H. E., & Williams, J. H. H. (2004). Measurement of Bovine IgG by Indirect Competitive ELISA as a Means of Detecting Milk Adulteration. *Journal of Dairy Science,* 87(3), 543-549.

Janstova, B., Drackova, M., Navratilova, P., Hadra, L., & Vorlova, L. (2007). Freezing point of raw and heat-treated goat milk. *Czech Journal of Animal Science*, 52(11), 394-398.

Jouppila, K., & Roos, Y. H. (1994). Glass Transitions and Crystallization in Milk Powders. *Journal of Dairy Science*, 77(10), 2907-2915.

Kawas, J. R., Lopes, J., Danelon, D. L., & Lu, C. D. (1991). Influence of forage-to-concentrate ratios on intake, digestibility, chewing and milk production of dairy goats. *Small Ruminant Research*, 4(1), 11-18.

Kelly P. M., J. I. G. J. S. (1989). Direct 'low-NOX' gas combustion heating of a spray drier during milk powder manufacture. *International Journal of Dairy Technology*, 42(1), 14-18.

Lu, C., Wang, G. R., Li, Y., & Zhang, L. B. (2013). Effects of homogenisation pressures on physicochemical changes in different layers of ultra-high temperature whole milk during storage. *International Journal of Dairy Technology*, 66(3), 325-332.

Manning, P. B., Coulter, S. T., & Jenness, R. (1968). Determination of Nitrate and Nitrite in Milk and Dry Milk Products1. *Journal of Dairy Science*, 51(11), 1725-1730.

Marshall, V., & El-Bagoury, E. (1986). Use of ultrafiltration and reverse osmosis to improve goats' milk yogurt. *International Journal of Dairy Technology*, 39(2), 65-66.

McLaughlin, F. (2006). *A brief comparison of United States and European Union standards for fluid dairy production*. A term paper for ANR 811.

Mele, M., Serra, A., Rafanelli, M. R., Conte, G., & Secchiari, P. (2005). Effect of forage/concentrate ratio and soybean oil supplementation on milk yield and quality from dairy goats. *Italian Journal of Animal Science*, 4(sup2), 392-394.

Morand-Fehr, P., & Sauvant, D. (1980). Composition and Yield of Goat Milk as Affected by Nutritional Manipulation1. *Journal of Dairy Science*, 63(10), 1671-1680.

Morand-Fehr, P., Fedele, V., Decandia, M., & Le Frileux, Y. (2007). Influence of farming and feeding systems on composition and quality of goat and sheep milk. *Small Ruminant Research*, 68(1-2), 20-34.

Murphy, B. P., Murphy, M., Buckley, J. F., Gilroy, D., Rowe, M. T., McCleery, D., & Fanning, S. (2005). In-line milk filter analysis: Escherichia coli O157 surveillance of milk production holdings.

International Journal of Hygiene and Environmental Health, 208(5), 407-413.

Nair, M. G., Mistry, V. V., & Oommen, B. S. (2000). Yield and functionality of Cheddar cheese as influenced by homogenization of cream. *International Dairy Journal*, 10(9), 647-657.

Park, Y. & Guo M. (2006). Goat milk products: processing technology, types and consumption trends. *Handbook of Milk of Non-Bovine Mammals.* Blackwell Publishers: 59-106. Iowa and Oxford, England.

Park, Y. M. (2010). Improving goat milk. Improving the safety and quality of milk. M. W. Griffiths. Cambridge, UK, Woodhead Publishing Limited: 304-346.

Park, Y. W. (2017). Goat milk–chemistry and nutrition. *Handbook of milk of non-bovine mammals*: 42-83.

Peter, H. D. (2000). *European systems for the safe production of raw cheese.* A report presented to the Vermont Cheese Council.

Pirisi, A., Lauret, A., & Dubeuf, J. P. (2007). Basic and incentive payments for goat and sheep milk in relation to quality. *Small Ruminant Research*, 68(1), 167-178.

Poutrel, B., & de Cremoux, R. (1997). Control of intramammary infections in goats: Impact on somatic cell counts. *Journal of Animal Science*, 75(2), 566.

Poutrel, B., & Lerondelle, C. (1983). Cell Content of Goat Milk: California Mastitis Test, Coulter Counter, and Fossomatic for Predicting Half Infection. *Journal of Dairy Science*, 66(12), 2575-2579.

Pulina, G. P. A. N. G. B. S. F. A. H. D. F. A. C. G. (2008). Nutrition and quality of goat's milk. *Dairy goats feeding and nutrition*, 1-30.

Raynal-Ljutovac, K., Gaborit, P., & Lauret, A. (2005). The relationship between quality criteria of goat milk, its technological properties and the quality of the final products. *Small Ruminant Research*, 60(1), 167-177.

Rodrigues, N. P. A., Givisiez, P. E. N., Queiroga, R. C. R. E., Azevedo, P. S., Gebreyes, W. A., & Oliveira, C. J. B. (2012). Milk adulteration: Detection of bovine milk in bulk goat milk produced by smallholders in northeastern Brazil by a duplex PCR assay. *Journal of Dairy Science*, 95(5), 2749-2752.

Santini, F. J., Lu, C. D., Potchoiba, M. J., Fernandez, J. M., & Coleman, S. W. (1992). Dietary fiber and milk yield, mastication, digestion, and rate of passage in goats fed Alfalfa Hay1, 2. *Journal of Dairy Science*, 75(1), 209-219.

Sanz Sampelayo, M. R., Chilliard, Y., Schmidely, P., & Boza, J. (2007). Influence of type of diet on the fat constituents of goat and sheep milk. *Small Ruminant Research*, 68(1), 42-63.

Schmidely, P. (2002). Lactation response and nitrogen, calcium, and phosphorus utilization of dairy goats differing by the genotype for αs1-casein in milk, and fed diets varying in crude protein concentration. *Journal of Dairy Science*, 85(9), 2299-2307.

Serradilla, J. M. (2001). Use of high yielding goat breeds for milk production. *Livestock Production Science*, 71(1), 59-73.

Ustunol, Z., & Brown, R. J. (1985). Effects of heat treatment and posttreatment holding time on rennet clotting of milk1. *Journal of Dairy Science*, 68(3), 526-530.

Van den Berg, G., C. Daamen. (1989). Bactofugation of cheese milk. N. *Eur. Food Dairy*, 55, 63-68.

Varnam, A. & Sutherland, J. P. (2001). *Cheese. Milk and Milk Products-Technology, Chemistry and Microbiology*. Springer Science & Business Media.

Wieking, W. (2004). *Removal of Somatic Cells from Raw Milk by State-of-the-Art Centrifugal Technology Bulletin of the International Dairy Federation* (389), 45-47.

Wszolek, R. K. R. A. Y. T. M. (2005). Microbiology of Fermented Milks. Dairy Microbiology *Handbook: The Microbiology of Milk and Milk Products*, Third Edition, 367-430.

Zervas, G., & Tsiplakou, E. (2011). The effect of feeding systems on the characteristics of products from small ruminants. *Small Ruminant Research*, 101(1), 140-149.

In: Goat Milk Chemistry …
Editor: Mingruo Guo
ISBN: 978-1-53617-299-7

Chapter 8

GOAT MILK NUTRITION AND HEALTH BENEFITS

Adam Killpartrick[1] and Mingruo Guo[2,3,*]
[1]Food Science Corporation, Inc. Williston, Vermont, US
[2]Key Laboratory of Dairy Science, Northeast Agricultural University, Ministry of Education, Harbin, Heilongjiang, China
[3]Department of Nutrition and Food Sciences, College of Agriculture and Life Sciences, University of Vermont, Burlington, Vermont, US

ABSTRACT

Goat's milk is a rich source of a variety of nutrients and can be considered, according to the European Concerted Action on Science of Functional Foods, to be a functional food. The digestibility of goat's milk due to its low content of lactose as well as the smaller size of the fat globules positions it as a better alternative to cow's milk. As such, it serves as an appropriate delivery system for infant formula and/or whey protein powders. There is also an inherent health benefit, particularly in the area of the gut and brain, to the dominant type of casein, A2 β-casein, found in

* Corresponding Author's Email: mguo@uvm.edu.

goat's milk. Further, the broad array of vitamins, minerals and fatty acids lend to its application to a variety of clinical conditions; and the prebiotic content, which help *Bifidobacteria* to thrive, is also important not just to gut health, but to immunity, neurological and cardio metabolic health. The studies examining goat's milk and the positive impact it has on the cardiovascular system, liver protection, immunity, anemia, osteoporosis and neurological health are compelling and warrant additional research.

Keywords: goat milk, nutrition, health benefits

Introduction

Due to its unique biochemical make-up goat milk has demonstrated that it has the capacity to provide significant health benefits where other milks, such as cow's milk and even human milk, have fallen short.

Specific nutritive compounds inherent to goat's milk, such as the vitamin and mineral content as well as the amino acid, lipid and protein profiles have positioned this widely available and sustainable form of milk as a potentially effective adjunctive to therapeutic interventions for the most common chronic diseases that our current population faces. And while the clinical research is still largely in its infancy, there have been promising results yielded relative to cardiovascular, neurological and digestive health to name a few.

This chapter will discuss the latest and most compelling research that has been conducted relative to these conditions as well as serve to substantiate the clinical applications of goat milk.

Goat Milk as a Functional Food

Functional foods are loosely defined as foods that have the potential to provide a positive effect on one's health beyond basic nutrition. According to the Consensus Document issued by the European Concerted Action on Science of Functional Foods, a food may be referred to as "functional," if it has been unequivocally proven that it positively influences one or more biological functions in the human body, improving the state of health and wellness, and reducing the risk to develop a disease (Diplock et al., 1999). With that definition in mind, and the forthcoming exploration into the research confirming the clinically impact of goat milk consumption, it is plain to see how many classify goat milk as a functional (or nutraceutical) drink.

DIGESTABILITY AND DISEASE

The digestibility of goat milk is due, at least in part, to the size of the fat globules of the goat milk, which are smaller than cow milk, therefore allowing it to be more easily digested. The composition of the protein of the goat milk allows forming a softer curd which may be easy to digest.

Goat milk does contain slightly lower level of lactose than that in cow milk. Lactose intolerance, as we know, is driven by a deficiency of lactase, the enzyme responsible for digesting milk sugar and lactose. In patients suffering from lactose intolerance, un-hydrolyzed lactose passes to large intestine where fermentation takes place resulting in the release of gases and other compounds causing gastrointestinal dysfunction, including, but not limited to, diarrhea, abdominal pain and flatulence (Russell, Ross, Fitzgerald, & Stanton, 2011). Lack of αs-casein also allows lactose to more rapidly pass through large intestine, hindering any fermentation process and thus preventing the symptoms of lactose intolerance (Robinson, 2001).

In terms of milk proteins, it is known that αs1-casein forms hard curds in the stomach, which might cause digestion problems in infants, but its concentration in goat milk is markedly lower than that in bovine milks (Bernacka, 2011).

In fact, the composition of goat milk protein is similar to that of breast milk and its casein fraction is mostly comprised of β-casein (Bernacka, 2011). Interestingly, β-casein in goat milk exists mainly as the A2 type rather than A1 beta casein (Cieślińska et al., 2012). A2 also does not produce β-casomorphin-7 (BCM-7), a bioactive opioid peptide and morphine-related compound generated during the process of milk digestion, which is theorized to be related with various disorders, such as gastrointestinal disturbances (Ho, Woodford, Kukuljan, & Pal, 2014) and even neurological conditions such as autism and schizophrenia (Kaminski, Cieslinska, & Kostyra, 2007).

And when considering substitutions for breast feeding, consider the hydrolysis of goat milk casein in the stomach is 96% compared to cow milk's 76-90% while human casein is completely hydrolyzed (Jasińska, 1995). The digestibility can be accessed via the time it takes to digest as well. Goat milk will typically break down in a baby's stomach in roughly 20 minutes. Cow's milk, on the other hand, may take up to 2-3 hours (Attaie, & Richter, 2000).

Nutrient Profile of Goat Milk

Vitamins

Specific vitamins that have been shown to have higher levels in goat's milk as compare to cow's milk include: vitamins A, D, B1, B2, B3 (by 3x) B6 and C. When it comes to immune health, vitamins A, C and D are vital. In fact, with the rampant deficiency of vitamin D across the US, it would be safe to say that a functional food so rich in vitamin D would be beneficial for a wide range of clinical conditions. Vitamin C is also critical for iron absorption; iron deficiency has emerged as a growing concern, and potential underlying cause, relative to neurological conditions such as ADHD, autism and insomnia.

When it comes to the B vitamins, we can simply look at the role each plays in energy production and the intimate connection to various systems the citric acid and mitochondrial health play in overall health. Not to mention the role of vitamin B6 in homocysteine metabolism, neurotransmitter health and diabetic neuropathy just to name a few.

Minerals

Minerals that are found at higher in goat's milk as compare to cow's milk include: Calcium, Magnesium, Potassium, Manganese and Selenium. Calcium is most notably connected to bone health but also plays a key role in regulating cardiovascular function, nerve transmission, muscular contractility and enzyme function. Magnesium is as important, if not more so, on account that it is involved in hundreds of vital reactions throughout the entire body. And as important as this mineral is to overall health, the Standard American Diet lacks adequate levels for optimal physiological function. From a clinical standpoint, the most prevalent dysfunction related to inadequate magnesium levels include hypertension, insomnia, anxiety and muscle tension. Ironically, certain medications such as diuretics (water pills used to treat hypertension) deplete the body of magnesium. Potassium is critical for cardiovascular health but deficiency has also been associated with arthritis, cancer and infertility. From a clinical standpoint, manganese is less frequently called out as compared to calcium or magnesium but plays a significant role in one of the most important protectors of the mitochondria, manganese superoxide dismutase (Holley, Bakthavatchalu, Velez-Roman, & Clair, 2011). And finally, selenium, also a key antioxidant; but a very important nutrient needed for the conversion of T4 thyroid hormone to the far more powerful and clinically relevant T3 hormone. With a hypothyroid case, presenting with blood T4 levels that are measured within range and T3 measuring low, there is typically a need for selenium as a first course of action as opposed to jumping to a medication.

Fatty Acids

In terms of lipids, goat's milk has 30-35% medium-chain fatty acids (C6-C14) MCT. This is compared to cow milk, which has only 15-20% (University of Granada, 2011). Medium chain triglycerides have emerged in recent years as a critical component for cardiovascular health, energy production via their unique ability to cross the double mitochondrial membrane very rapidly, and do not require the presence of carnitine in order to stimulate energy production.

Additionally, goat milk is rich in Omega 6 and omega 3, especially EPA and DHA (Li et al., 2017). From a clinical perspective EPA is largely viewed as the 'anti-inflammatory' component. DHA, on the other hand, is considered to be vital for brain and eye health in the elderly population as well as neural development in babies. This is why there is such an emphasis on DHA for babies and the rationale behind the inclusion of goat milk in infant formula.

Another fatty acid, decanoic acid, is particularly relevant in the current keto-climate. Decanoic acid provides a direct inhibition of excitatory neurotransmission in the brain which contributes to the anticonvulsant effect of the MCT ketogenic diet frequently recommended to epileptic patients (Chang et al., 2016).

And then there is Lauric Acid, shown to have an impact on HDL levels. However, despite a connection to HDL, the correlation to a decrease in cardiovascular disease risk is uncertain. Lauric acid is widely known as the most antimicrobial of the saturated fatty acids as well. Intradermal injections and topical application has demonstrated the antimicrobial capacity to decrease the ropionibacterium acnes (P. acnes), which promotes follicular inflammation (inflammatory acne) and is considered an alternative to antibiotic treatment of acne vulgarus (Nakatsuji et al., 2009).

Proteins and Amino Acids

Goat milk provides roughly 9 grams of protein per cup, which exceeds the roughly 8 grams provided by cow's milk. USDA's national nutrient database contains18 amino acids of which goat milk contains higher amounts of 14 of them as compared to cow's milk.

Proteins of the goat milk are important source of the angiotensin converting enzyme (ACE) inhibitory peptides. From a clinical perspective, it has been found that ACE activities in the aorta, left ventricle, and kidney were significantly decreased in the goat milk supplemented group compared with those of the control group (Geerlings et al., 2006). This serves as a novel and well documented approach to supporting cardiovascular physiology.

Carnitine, which is also abundantly found in goat's milk, also provides key support for energy production (via the citric acid cycle), heart function and also brain function (Ulusoy, 2015).

Probiotic and Prebiotics

The importance of prebiotics and probiotics on digestive health, as skin, endocrine and musculoskeletal systems, the gut brain axis and vitamin synthesis (as it relates to B vitamins), cannot be overstated.

As mentioned, goat milk has higher level of the oligosaccharides as human milk than cow milk. It is well known that, these serve as prebiotics in gut and have the potential to support the health of the gastrointestinal tract overall (Raynal-Ljutovac, Lagriffoul, Paccard, Guillet, & Chilliard, 2008). Oligosaccharides are responsible for the helping key beneficial bacteria thrive i.e., Bifidobacteria. Bifidobacteria species exert a broad array of health benefits including immunomodulation, supporting healthy cholesterol levels and improving lactose digestion (Russell et al., 2011).

Clinical trials have shown that various oligosaccharides can be used to stimulate bifidobacteria in the GI- tract. These include inulin, fructo oligosaccharides (FOS), galactooligosaccharide and lactulose.

Studies examining oligosaccharides from goat's milk have shown, when included as a constituent in infant formula, have the potential to stimulate host bifidobacteria levels to increase. In fact, the increase is similar to levels found in the digestive systems of breast-fed babies. Goats' milk oligosaccharides, particularly 6-sialyl lactose, constitute the "soluble fiber" which provides additional nutrients for colonic bacteria. These oligosaccharides set the stage for bacteria colony forming to thrive, specifically by providing the substrate necessary for that to occur (Kim, Yun, Oh, & Yoon, 2015). Goat milk oligosaccharides have been shown to exhibit significant intestinal anti-inflammatory effects in experimental models of mouse colitis (Lara-Villostada et al., 2006). This finding actually lead to a study that looked at goat whey protein's impact on the intestinal inflammatory process. Findings mirrors those of earlier findings; goat whey was able to mitigate the evaluated clinical signs and inhibit the secretion of pro-inflammatory cytokines such as IL-1β, IL-6, IL-17 and TNF-α. It was also found that 'a reduction in inflammation was also evidenced by a decrease in the microscopic damage score of the colonic tissue' (Araújo et al., 2017).

CARDIOVASCULAR HEALTH

At the time of this writing, cardiovascular disease (CVD) was listed as the underlying cause of over 800,000 deaths in the United States. According to the American Heart Association's 2018 Statistics, coronary heart disease and stroke were the leading causes of CVD, accounting for 43.8% and 16.8% of the deaths respectively. Both coronary heart disease and ischemic stroke are driven by oxidation of cholesterol and plaguing along the arterial walls. In terms of the types of cholesterols, low-density lipoprotein (LDL), an atherogenic lipoprotein which transports cholesterol from the liver to the blood vessels and is often called "the bad cholesterol." Incidentally, 1 in 3 Americans, the same number of people who die from heart disease, have elevated LDL. High-density lipoprotein (HDL) is widely considered the 'good cholesterol' as it has demonstrated a beneficial effect on platelet function, endothelial function, coagulation parameters, inflammation, and interactions with triglyceride-rich lipoproteins.

Goat milk was shown in an animal study to enhance HDL levels leaving researchers to conclude that goat milk does, in fact, protect against heart disease (Rodrigues et al., 2014). Other studies have also revealed a hypocholesterolemia effect, in which goat milk consumption reduced the level of plasma triglyceride and exerted a positive effect on lipid metabolism (López-Aliaga et al., 2005).

Goat milk is rich in monounsaturated fatty acids (MUFA), polyunsaturated fatty acids (PUFA) and medium chain triglycerides (MCT), all of which are beneficial for the cardiovascular conditions. Additionally, it consistently demonstrates lower level of the cholesterol than cow milk (Haenlein, 2004). Goat milk exerts hypocholesterolemia effect. Goat milk consumption reduces plasma triglyceride and hence had positive effect on lipid metabolism (López-Aliaga et al., 2005).

Goat milk may also play a role in vasodilation directly as well. An in vitro study has shown cells isolated from humans who had been drinking goat milk, from different breeds, were triggered to release nitric oxide (NO). In turn, the NO reaches the blood stream via the lymphatic system, thus promoting vasodilation and exerting a cardio protective and anti-arthrogenic affect (Park, Juarez, Ramos & Haenlein, 2007).

HEPATOPROTECTION

Liver disease, specifically alcoholic liver disease (ALD) and nonalcoholic fatty liver disease (NAFLD), have both steadily increased in the past decade. Age-standardized mortality increased for ALD (2.3% from 2007 through 2013 and more than doubled to 5.5% from 2013 through 2016) and nonalcoholic fatty liver disease 6.1% from 2007 through 2013 to 11.3% from 2013 through 2016) (Kim et al., 2018). While goat milk certainly cannot prevent ALD or NAFLD, it may be able to provide a level of protection.

One study aimed to assess whether goat milk did, in fact, harbor the potential to provide any such protection when faced with hepatotoxicity induced by antitubercular drugs in rats.

Hepatotoxicity was induced via a combination of rifampicin, isoniazid and pyrazinamide given orally for 30 days. Treatment groups received goat milk along with antitubercular drugs. Liver damage was assessed using biochemical and histological parameters.

Goat milk (20 mL/kg and 40 mL/kg) was effective in reversing the rise in malondialdehyde level, often used as a biomarker for excessive oxidative stress in cancer patients. There was also a significant decrease in serum alanine aminotransferase and aspartate aminotransferase in the goat milk group as compared with the antitubercular drug suspension groups. In addition to preventing damage, another significant finding was that both goat milk groups prevented a decline in superoxide dismutase, which is expected to fall in when exposed to such toxic conditions.

Histological examination indicated that goat milk reduced inflammation and necrotic changes in hepatocytes in the treatment groups. The results indicated that goat milk prevented the antitubercular drug-induced hepatotoxicity and is an effective hepatoprotective agent (Miglani, Patyar, Patyar, & Reshi, 2016).

IMMUNOLOGY AND ALLERGY SUPPORT

It has been proven that goat milk is beneficial from an immunological standpoint, specifically for reducing specific markers involved in the allergic response. Compared to goat milk, cow milk significantly increased various bio-markers of inflammation, including but not limited to cytokine interleukin-4 (IL-4) and antigen-specific immunoglobulin G1 (IgG1), key markers in hypersensitivity reactions. IgG1 binds to mast cells and imitates a cascade effect which results in degranulation (the ultimate response to an allergic response), which releases histamine and results in an allergic reaction. This is not found to be the case with goat's milk, which did not induce an allergic response (Lara-Villoslada, Olivares, Jimenez, Boza, & Xaus, 2004).

Similarly, another study found that drinking of cow milk had significantly higher levels of the inflammatory marker tumor necrosis factor-α (TNF-α) than those who consumed goat milk. TNF-α is vitally important and a primary mediator of adverse reactions to cow milk protein including gastrointestinal distress, respiratory and skin conditions such as eczema. Not only were there no inflammatory cytokines notes among goat milk consumption subjects, those who consumed goat milk had higher levels of the cytokine IL-10, which is known as a potent anti-inflammatory bio-marker (Albenzio et al., 2012).

Goat milk contains a high level of conjugated linoleic acid (CLA) (Jirillo et al., 2010). The anti-carcinogenic potential of CLA has been reported in cases of colon cancer (Dashwood, 1995) in animal models and in vitro melanoma models(Shultz, Chew, Seaman, & Luedecke, 1992), breast and colorectal cancer (Durgam & Fernandes, 1997). CLA's mechanism of action is thought to lie in its ability to inhibit tumor proliferation, however, it is not fully understood. It is thought, though, that deviation from the eicosanoid-dependent cell signaling systems, oxidative effects and disturbance of the receptor mediated actions of estrogen are all potential drivers (Jirillo et al., 2010).

OSTEO SUPPORT

Currently it is estimated that roughly 200 million people worldwide suffer from osteoporosis. This is complicated by shifts in hormones, as in the case of menopause; approximately 30% of postmenopausal women are diagnosed with osteoporosis in the US and in Europe. A large percentage of these women, roughly 40%, will exhibit some degree of fragility fractures.

Taking this information into consideration, it is a common approach when presented with any degree of bone loss (osteopenia or osteoporosis) that an increased consumption of milk should take place with the belief in its benefits based predominately on its calcium content. While this isn't entirely accurate, the thought process is leaning in the right direction, especially if goat milk is the type of milk being consumed.

When compared to cow milk, there are consistently higher levels of calcium and vitamin D found in goat milk which provide added support for healthier bones. However, in the increase in bioavailability of zinc and selenium in goat milk (Barrionuevo et al., 2003) zinc has demonstrated a role in supporting the mineralization and bone structure as well for patient's osteopenia or osteoporosis (Yamaguchi, 2010). Selenium, one of the bodies' master antioxidants also lends itself to the integrity of the bone's structure. This is a direct result of the role reactive oxygen species (ROS) play in osteoporosis; ROS induce the apoptosis of osteoblasts and osteocytes, and this conversely supports osteoclast genesis and inhibits mineralization and osteogenesis (Domazetovic, Marcucci, Iantomasi, Brandi, & Vincenzini, 2017).

Another novel area of interest in terms of goat milk protein is that of the CSN1S2 protein. This protein delivered in goat milk was found to repair femur crystallization and mesostructure in CFA-induced rheumatoid arthritis in rats as compared to placebo.

The CSN1S2 protein also had a reparative effect on the collagen structure of the trabecular bone, which was indicated by the reduction in holes and the rearrangement of fibrillar collagen into step-ladder patterns (smooth surfaces) (Fatchiyah, Setiawan, Suharjono, & Noor, 2015).

IRON, ANEMIA AND ASSOCIATED CONDITIONS

Iron deficiency is the most globally prevalent nutritional problem reaching an epidemic level in many developing countries. During pregnancy, iron deficiency is the leading cause of anemia in infants and young children.

One of the most common neurological conditions in the pediatric population is Attention Deficit Hyperactivity Disorder (ADHD). It has been found that low serum iron, ferritin levels, and vitamin D deficiency may be associated with ADHD (Bener, 2015). Based on the content of both bioavailable iron and vitamin D found in goat milk, it would stand to reason that increasing dietary intake of goat milk may positively impact children with ADHD.

High prevalence of iron deficiency (ID) and iron deficiency anemia (IDA) has also been reported in children, not as a causative factor but concurrently, with autism spectrum disorder (ASD). In fact, 200 subjects between the ages of 2 and 18 were evaluated via serum ferritin as the indicator for iron deficiency. Hemoglobin, hematocrit, iron, MCV and RDW were also measured. The presence of iron deficiency anemia in ASD patients was double that of the controls. It was also found that IDA was significantly higher in children with severe ASD (Gunes, Ekinci, & Celik, 2017). Another study from a decade earlier looked at the effect of iron supplementation in patients with ASD and found that 76% of those with ASD who also had trouble sleeping experienced significant improvement in their sleep patterns (Dosman et al., 2007). This is encouraging and indicative that more research is needed looking at the impact of increasing iron intake in the ASD population.

Iron deficiency connection to insomnia is not strictly relegated to patients with ASD. Due to the key role of iron in the metabolism of monoamines in the brain and the role of those very same monoamines in sleep physiology, it was found that sleep quality did, in fact, decline in cases of IDA (Murat et al., 2015).

The health ramifications, especially for young children who are already fighting an uphill battle against neurological conditions such as ADHD, autism and insomnia, are significant. Fortunately, the increased bioavailability of iron (Park, Mahoney, & Hendricks, 1986) that goat milk provides can contribute to the restoration and maintenance of healthy blood iron levels.

NEUROLOGY

Anxiety disorders are the most common mental illness in the U.S., affecting 40 million adults in the United States age 18 and older, or 18.1% of the population every year. Medications have long been the standard course of treatment for these conditions, however, nutritional intervention has emerged in recent years, and continues to do so, as a primary part of an integrative medicine approach to anxiety disorders.

One promising animal study looked a diet with goat milk fat supplemented with exercise on oxidative stress bio-markers in the brain and anxiety behavior in an animal model. Adult male rats were randomly divided into six groups, each comprised of 10 subjects:(n = 10 per group): control sedentary (CON) and exercised (CON-EX) groups, goat milk fat sedentary (GM) and exercised (GM-EX) groups, goat milk fat enriched with conjugated linoleic acid (CLA) sedentary (GM-CLA) and exercised (GM-CLA-EX) groups.

The exercised groups were submitted to treadmill running during a period of four weeks. Anxiety parameters were evaluated as were Malondialdehyde (MDA) and glutathione (GSH) levels in the brain. Compared to the CON and GM groups, the GM-CLA group, where the animals were administered goat milk fat with a supplemented amount of conjugated linoleic acid (CLA), showed a greater response. As was expected, all exercised animals demonstrated an increase in grooming when compared to sedentary animals and MDA levels in the brain were lower in the GM group when compared to those in the CON group. GM and GM-CLA groups showed a significant elevation in the glutathione (GSH) levels in the brain.

The clinical applications of this alone are significant. GSH has long been considered the 'master antioxidant' and been implicated in such disease processes such as cancer (Traverso et al., 2013), Alzheimer's (Saharan & Mandal, 2014), possibly Parkinson's (Smeyne & Smeyne, 2013).

Goat milk fat also showed a potential effect on the reduction of oxidative stress, which was associated with a reduction in anxiety behavior. However, this beneficial effect was reversed when exercise was involved. This is expected as studies show that acute exercise induces reactive oxygen and nitrogen species and, subsequently, oxidative stress; whereas regular exercise training induces the endogenous antioxidative system and protects the body against adverse effects of oxidative damage (Muaz Belviranli, 2006). This decrease in oxidative stress coincided with a decrease in anxiety as well.

REFERENCES

Albenzio, M., Campanozzi, A., D'Apolito, M., Santillo, A., Mantovani, M. P., & Sevi, A. (2012). Differences in protein fraction from goat and cow milk and their role on cytokine production in children with cow's milk protein allergy. *Small Ruminant Research*, 105(1), 202-205.

Araújo, D. F. S., Guerra, G. C. B., Pintado, M. M. E., Sousa, Y. R. F., Algieri, F., Rodriguez-Nogales, A., ... Rodriguez-Cabezas, M. E. (2017). Intestinal anti-inflammatory effects of goat whey on DNBS-induced colitis in mice. *PLoS One*, 12(9), 1-19.

Attaie, R., & Richter, R. L. (2000). Size distribution of fat globules in goat milk. *Journal of Dairy Science*, 83(5), 940-944.

Barrionuevo, M., Aliaga, I. L., Alferez, M. J. M., Mesa, E., Nestares, T., & Campos, M. S. (2003). Beneficial effect of goat milk on bioavailability of copper, zinc and selenium in rats. *Journal of Physiology and Biochemistry*, 59(2), 111-118.

Bener, A. (2015). Higher Prevalence of Iron Deficiency as Strong Predictor of Attention Deficit Hyperactivity Disorder in Children. *European Psychiatry*, 30, 691.

Bernacka, H. (2011). Health-promoting properties of goat milk. *Medycyna Weterynaryjna*, 67(8), 507-511.

Chang, P., Augustin, K., Boddum, K., Williams, S., Sun, M., Terschak, J. A., Williams, R. S. B. (2016). Seizure control by decanoic acid through direct AMPA receptor inhibition. *Brain*, 139, 431-443.

Cieślińska, A., Kostyra, E., Kostyra, H., Oleński, K., Fiedorowicz, E., & Kamiński, S. (2012). Milk from cows of different β-casein genotypes as a source of β-casomorphin-7. *International Journal of Food Sciences & Nutrition*, 63(4), 426-430.

Diplock, A. T., Aggett, P. J., Ashwell, M., Bornet, F., Fern, E. B., & Roberfroid, M. B. (1999). Scientific concepts of functional foods in Europe consensus document. *British Journal of Nutrition*, 81(4), S1-S27.

Domazetovic, V., Marcucci, G., Iantomasi, T., Brandi, M. L., & Vincenzini, M. T. (2017). Oxidative stress in bone remodeling: role of antioxidants. *Clinical Cases in Mineral and Bone Metabolism*, 14(2), 209-216.

Dosman, C. F., Brian, J. A., Drmic, I. E., Senthilselvan, A., Harford, M. M., Smith, R. W., Roberts, S. W. (2007). Children with Autism: Effect of iron supplementation on sleep and ferritin. *Pediatric Neurology*, 36(3), 152-158.

Durgam, V. R., & Fernandes, G. (1997). The growth inhibitory effect of conjugated linoleic acid on MCF-7 cells is related to estrogen response system. *Cancer Letters*, *116*(2), 121-130.

Fatchiyah, F., Setiawan, B., Suharjono, S., & Noor, Z. (2015). The anti-osteoporosis effects of CSN1S2 protein of goat milk and yoghurt on a complete Freund's adjuvant-induced rheumatoid arthritis model in rats. *Biomarkers and Genomic Medicine*, 7(4), 139-146.

Geerlings, A., Villar, I. C., Hidalgo Zarco, F., Sánchez, M., Vera, R., Zafra Gomez, A. & Duarte, J. (2006). Identification and Characterization of Novel Angiotensin-Converting Enzyme Inhibitors Obtained from Goat Milk. *Journal of Dairy Science*, 89(9), 3326-3335.

Gunes, S., Ekinci, O., & Celik, T. (2017). Iron deficiency parameters in autism spectrum disorder: clinical correlates and associated factors. *Italian Journal of Pediatrics*, 43, 6.

Haenlein, G. F. W. (2004). Goat milk in human nutrition. *Small Ruminant Research*, 51(2), 155-163.

Holley, A. K., Bakthavatchalu, V., Velez-Roman, J. M. & Clair, K. S. (2011). Manganese Superoxide Dismutase: Guardian of the Powerhouse. *International Journal of Molecular Sciences*. 12(10), 7114-62.

Ho, S., Woodford, K., Kukuljan, S., & Pal, S. (2014). Comparative effects of A1 versus A2 beta-casein on gastrointestinal measures: a blinded randomised cross-over pilot study. *European Journal of Clinical Nutrition*, 68(9), 994-1000.

Jasińska, B. (1995). The comparison of pepsin and trypsin action on goat, cow, mare and human caseins. *Roczniki Akademii Medycznej W Bialymstoku*, 40(3), 486-493.

Jirillo, F., Martemucci, G., D'Alessandro, A. G., Panaro, M. A., Cianciulli, A., Superbo, M., Magrone, T. (2010). Ability of goat milk to modulate healthy human peripheral blood lymphomonocyte and polymorphonuclear cell function: *in vitro* effects and clinical implications. *Current Pharmaceutical Design*, 16(7), 870-876.

Kaminski, S., Cieslinska, A., & Kostyra, E. (2007). Polymorphism of bovine beta-casein and its potential effect on human health. *Journal of Applied Genetics*, 48(3), 189-198.

Kim, D., Li, A. A., Gadiparthi, C., Khan, M. A., Cholankeril, G., Glenn, J. S., & Ahmed, A. (2018). Changing trends in etiology-based annual mortality from chronic liver disease, from 2007 through 2016. *Gastroenterology*, 155(4), 1154-1163.e1153.

Kim, H. H., Yun, S. S., Oh, C. H., & Yoon, S. S. (2015). Galactooligo-saccharide and Sialyllactose content in commercial lactose powders from goat and cow milk. *Korean Journal for Food Science of Animal Resources*, 35(4), 572-576.

Lara-Villoslada, F., Olivares, M., Jimenez, J., Boza, J., & Xaus, J. (2004). Goat milk is less immunogenic than cow milk in a murine model of atopy. *Journal of Pediatric Gastroenterology and Nutrition*, 39(4), 354-360.

Lara-Villostada, F., Debras, E., Nieto, A., Concha, A., Galvez, J., Lopez-Huertas, E., Xaus, J. (2006). Oligosaccharides isolated from goat milk reduce intestinal inflammation in a rat model of dextran sodium sulfate-induced colitis. *Clinical Nutrition*, 25(3), 477-488.

Li, Q., Zhao, Y., Zhu, D., Pang, X., Liu, Y., Frew, R., & Chen, G. (2017). Lipidomics profiling of goat milk, soymilk and bovine milk by UPLC-Q-Exactive Orbitrap Mass Spectrometry. *Food Chemistry*, 224, 302-309.

Liew, C., Schut, H. A. J., Chin, S. F., Pariza, M. W., & Dashwood, R. H. (1995). Protection of conjugated linoleic acids against 2-amino-3-methylimidazo[4,5-f] quinoline-induced colon carcinogenesis in the F344 rat: a study of inhibitory mechanisms. *Carcinogenesis*, *16*, 3037-3043.

López-Aliaga, I., Alférez, M. J. M., Nestares, M. T., Ros, P. B., Barrionuevo, M., Campos, M. S. (2005) Goat milk feeding causes an increase in biliary secretion of cholesterol and a decrease in plasma cholesterol levels in rats, *Journal of Dairy Science* (88), 1024-1030.

Miglani, S., Patyar, R. R., Patyar, S., & Reshi, M. R. (2016). Effect of goat milk on hepatotoxicity induced by antitubercular drugs in rats. *Journal of Food and Drug Analysis*, 24(4), 716-721.

Muaz Belviranli, H. G. (2006). Review-acute exercise induced oxidative stress and antioxidant changes. *European Journal of General Medicine*, 18(3).

Murat, S., Ali, U., Serdal, K., Suleyman, D., Ilknur, P., Mehmet, S., ... Tunahan, U. (2015). Assessment of subjective sleep quality in iron deficiency anaemia. *African Health Sciences*, 15(2), 621-626. doi:10. 4314/ahs.v15i2.40.

Mwenze P. (2015), *Research Journal of Agriculture and Environmental Management*. Vol. 4(9), 343-349.

Nakatsuji, T., Kao, M. C., Fang, J.-Y., Zouboulis, C. C., Zhang, L., Gallo, R. L., & Huang, C.-M. (2009). Antimicrobial property of lauric acid against propionibacterium acnes: Its therapeutic potential for inflammatory acne vulgaris. *Journal of Investigative Dermatology*, 129(10), 2480-2488.

Park, Y. W., Juarez, M., Ramos, M. & Haenlein, G. F. W. (2007). Physicochemical characteristics of goat and sheep milk. *Small Ruminant Research*. 68 (68), 88-113.

Park, Y. W., Mahoney, A. W., & Hendricks, D. G. (1986). Bioavailability of iron in goat milk compared with cow milk fed to Anemic rats. *Journal of Dairy Science*, 69(10), 2608-2615.

Raynal-Ljutovac, K., Lagriffoul, G., Paccard, P., Guillet, I., & Chilliard, Y. (2008). Composition of goat and sheep milk products: An update. *Small Ruminant Research*, 79(1), 57-72.

Robinson, F. (2001). Goats milk–a suitable hypoallergenic alternative? *British Food Journal*, 103(3), 198-208.

Rodrigues, R., Soares, J., Garcia, H., Nascimento, C., Medeiros, M., Bomfim, M., Queiroga, R. (2014). Goat milk fat naturally enriched with conjugated linoleic acid increased lipoproteins and reduced triacylglycerol in rats. *Molecules*, 19(3), 3820-3831.

Russell, D. A., Ross, R. P., Fitzgerald, G. F., & Stanton, C. (2011). Metabolic activities and probiotic potential of bifidobacteria. *International Journal of Food Microbiology*, 149(1), 88-105.

Saharan, S., & Mandal, P. K. (2014). The emerging role of glutathione in Alzheimer's disease. *Journal of Alzheimers Disease*, 40(3), 519-529.

Shultz, T. D., Chew, B. P., Seaman, W. R., & Luedecke, L. O. (1992). Inhibitory effect of conjugated dienoic derivatives of linoleic acid and β-carotene on the in vitro growth of human cancer cells. *Cancer Letters,* 63(2), 125-133.

Smeyne, M., & Smeyne, R. J. (2013). Glutathione metabolism and Parkinson's disease. *Free Radical Biology and Medicine*, 62, 13-25.

Traverso, N., Ricciarelli, R., Nitti, M., Marengo, B., Furfaro, A. L., Pronzato, M. A., Domenicotti, C. (2013). Role of glutathione in cancer progression and chemoresistance. *Oxidative Medicine and Cellular Longevity*, 2013, 972913-972913.

Ulusoy, B. H. (2015). Nutritional and health aspects of goat milk consumption. *Keçi Sütü Tüketiminin Beslenme ve Sağlık Üzerine Etkileri.*, 13(1), 56-60.

University of Granada. Goat milk can be considered as functional food, Spanish researchers find. *Science Daily*, 2011, May 19.

Yamaguchi, M. (2010). Role of nutritional zinc in the prevention of osteoporosis. *Molecular and Cellular Biochemistry*, 338(1-2), 241-254.

In: Goat Milk Chemistry …
Editor: Mingruo Guo
ISBN: 978-1-53617-299-7

Chapter 9

OUTLOOK OF GOAT MILK PRODUCTION, PRODUCTS AND MARKET

Mingruo Guo*

Department of Nutrition and Food Sciences, College of Agriculture and Life Sciences, University of Vermont, Burlington, Vermont, US
Key Laboratory of Dairy Science (Northeast Agricultural University), Ministry of Education, Harbin, Heilongjiang, China

DAIRY GOAT INDUSTRY OUTLOOK

Dairy goat industry will continue to grow for the foreseeable future. Goats are widely distributed around the world due to their adaptable nature and survival ability. More people drink goat milk than the milk of any other mammalian species although goat milk production just accounts less than 5% of the total milk production in the world. Dairy goat farming plays a very important role in the nutrition and economy for many developing countries. Goat milk and goat milk products consumption provide one of the stable

---

* Corresponding Author's Email: mguo@uvm.edu.

sources of calories, protein, calcium and other minerals for the people around the Middle East, Southern Asia, and some tropical countries. The investment for dairy goat farming operations is relatively small compared with dairy cow operations, and the return could be tremendous. Goat milk is also considered as a good alternative to bovine milk and recommended by health professionals for those who may be allergic to the latter and other concerns about cow milk.

Goat milk and its products are considered as specialties in North America and European countries. Because new processing technologies have been developed and are readily available, goat milk can be utilized to produce all the dairy products as those made from cow milk. The goat milk products market is growing and the demands are strong. Therefore, the future of dairy goat industry is bright and will be growing at a reasonable rate.

Current Challenges for Goat Milk Production, Product Safety, New Product Development

It is commonly agreed that the focus and investment for dairy goat milk production and product research are far less than that for dairy cow industry. Goats are different from cows in terms of physiology, behavior, body structure, nutrition requirements, and lactation properties. There are also differences in chemical composition, physiochemical properties, biological properties and processing technology apsects between bovine milk and caprine milk. In order to be profitable and sustainable for dairy goat operations and the goat milk products processors, milk production efficiency and management, product development and quality control, and specific processing technology for goat milk products need to be studied on and deserve more resources.

Milk Production Efficiency and Goat Herd Management

More than 90% of dairy goats are distributed in Asian and African countries, only 8% of total goat population is raised in Europe and North America. However, North America and Europe provide more than 16% of total goat milk supply for the world. It will make a big difference in terms of milk production if we could help the developing nations to improve their production efficiency and management systems. Improvement of milk production seasonality is also one of the areas that may need to be improved and managed.

Product Safety

Although dairy goats are resistant to certain diseases, safety and quality control of goat milk products should be one of the critical issues for product manufacturing and distribution management for fresh and raw milk cheese products, pasteurized milk, and products made in small processing facilities and in warm climate regions. Special attention must be paid to the new emergence of foodborne pathogens. Several infectious diseases are still threating to the farmers and milk handlers, and even to the milk product consumers in developing countries.

Goat Milk Flavor Property and Management

Goat milk and its products may have a unique flavor,often called goaty flavor. The goaty flavor may not be welcomed by certain segments of the population. There are efforts to improve or minimize the flavor impact on the consumption of goat milk products both from academia and dairy foods industry. This flavor issue could be managed by using genetic, biological, physical/technical, and chemical methods. A team of scientists in the editor's laboratory is working hard on this subject. Two patents have been awarded and published. Their findings are promising and will be reported in due course.

Biological Properties of Goat Milk

Goat milk and goat milk products have been used as alternatives to cow milk and its products according to consumers demands or professional advice, for example, goat milk is easier to be digested and may cause less allergic responses. However, there is very limited scientific evidence or sound results to back up these claims. Therefore, systematic and scientifically designed studies on these biological properties are needed to investigate if there are any advantages of consumption of goat milk and its products compared with its counterparts of cow milk.

The information about proteomics, lipidomics, and microbiome of goat milk are very limited. These topics should be systematically investigated since new discoveries of these important areas will provide foundations for further use of goat milk as a major ingredient for development of new products and especially for goat milk-based functional foods.

Goat Milk Product Processing Properties

It has been well documented that heat and alcohol stability of goat milk are lower than those of cow milk, especially the milk from small rural dairy goat operations and certain stages of lactation. It must be pointed out that caution should be excised when applying the processing technology of cow milk products for manufacturing goat milk products. The processing technology may need modifications. New research should be conducted before opening any large-scale goat milk processing operations.

It is hard to produce goat milk yogurt that may have the comparable consistency to cow milk's yogurt since the gel of goat milk is a rather weak and fluid-like structure. This is mainly due to the difference in chemistry of the caseins and casein micelle structure between goat milk and cow milk. The acid induced gelation properties of goat milk can be improved by physical, chemical and enzymatic means (see Chapter 6). Membrane technology should be exploited for improving goat milk gelation property and structure quality of related products.

Whey Utilization

The majority of cheese products made from goat milk are fresh cheeses such as Chevre and Greek yogurt-like products which may generate a large quantity of acid whey. Acid whey utilization is a big challenge for the industry due to high salt content and low pH value, which is difficult to make food grade whey powder or whey protein products. Special designed membrane technology similar to bovine milk whey processing could be developed to solve this problem. Demineralized goat milk whey is in high demand since it is one of the major ingredients for goat milk based infant formula formulation. The low ash content whey products are also in demand for value added products like pet foods.

New Product Development

Bottled milk, yogurt and cheese are main commercial goat milk products on the markets across the globe. Goat milk can be used as a raw material to produce almost all the varieties of dairy products like cow milk. Probiotic or symbiotic fermented goat milk products including kefir, yogurt and drinkable yogurt are getting very popular in many major markets. Infant formulas and formulated supplement products for pets will have their market shares. Goat milk whey products such as demineralized whey powder and whey protein products are all in high demand around the world, especially in the new emerging markets.

EDITOR'S CONTACT INFORMATION

Mingruo Guo

Department of Nutrition and Food Sciences,

College of Agriculture and Life Sciences,

University of Vermont, Burlington, Vermont, US

Email: mguo@uvm.edu

Fax: 1-802-656-0001

INDEX

B

C

D

G

H

I

J

K

L

Q

R

S

T

U

V

W

Y

Z

α

β

κ